INEVITABLE

EV

INEVITABLE

INSIDE THE MESSY, UNSTOPPABLE TRANSITION TO ELECTRIC VEHICLES

MIKE COLIAS

HARVARD BUSINESS REVIEW PRESS • BOSTON, MASSACHUSETTS

Copyright 2025 Mike Colias

All rights reserved

Printed in Great Britain by Bell and Bain Ltd, Glasgow

10 9 8 7 6 5 4 3 2 1

The web addresses referenced in this book were live and correct at the time of the book's publication but may be subject to change.

Cataloging-in-Publication data is forthcoming.

ISBN: 978-1-64782-538-6
eISBN: 978-1-64782-539-3

The paper used in this publication meets the requirements of the American National Standard for Permanence of Paper for Publications and Documents in Libraries and Archives Z39.48-1992.

For my wife, Stacy, and my parents.

CONTENTS

AUTHOR'S NOTE

When I started on the auto-industry beat in the fall of 2010, electric cars were part of the story even back then, but little more than a novelty. The Chevy Volt and Nissan Leaf were just being shipped to dealerships. Tesla's Model S, which would be hitting the road soon after, was a curiosity. No major automaker was looking to jump into EVs in a major way.

That began to change around 2017. Tesla's Elon Musk was straining to make the Model 3 a mass-market reality. Big legacy carmakers like General Motors and VW finally came around to the view that the future was electric. My job as an automotive reporter for the *Wall Street Journal* afforded me the ideal perch from which to report on this massive global pivot.

In researching and reporting this book, I drew on my work covering the EV story for the *Journal* over the years. Much of the book relies on dozens of additional interviews and reporting trips that provided the scenes, anecdotes, direct quotes, and dialogue to stitch together the narrative arc of this book. Many of them are based on firsthand accounts, and every effort was made to verify those by at least two people. Throughout the book, quotes unattributed to other sources are from my own interviews and reporting. I also have extensively used media coverage of the auto industry from many outlets, including respected competitors at Reuters, Bloomberg, *Automotive News*, CNBC and the *New York Times*. I took great care to credit their work both in the text and endnotes.

INEVITABLE

A Messy Reckoning

About a dozen Ford executives gathered inside a nearly empty warehouse on the automaker's campus in Dearborn, Michigan. They were there for an autopsy.

It was early 2022. After a late start, Ford had loudly announced its arrival in the electric-vehicle race. Its muscular Mustang Mach-E SUV was a relative hit, the top-selling US electric model without Tesla in its name. A battery-powered version of the company's top-selling F-150 pickup truck, the Lighting, had a waiting list stretching more than two years and was being plugged by Jimmy Fallon.

But Ford CEO Jim Farley was worried. Then in his late fifties, with a boyish mop of brown hair, Farley wanted to see this teardown of his own company's Mustang Mach-E right next to a teardown of a Tesla Model 3. It was an engineering reality check.

The brusque and intense CEO had long been fixated on Tesla and how Elon Musk's company had been able to accelerate so far ahead on EVs. Farley, a longtime marketing guy, isn't an engineer. But he is obsessed with cars and has a piercing curiosity and almost savant-like mastery when it comes to the details of what's under the hood. On this day, his race-car driver's mind was racing.

Ford's designers and engineers had rushed its early EVs to market by retrofitting their gas-powered cousins. The Mach-E and Lightning were essentially placeholders, just to get Ford in the EV game. The company already was at work on a next-generation of electrics designed from the ground up to be powered by batteries and motors, rather than internal-combustion engines. Those future EVs promised longer driving ranges, faster charging times, roomier interiors, cooler looks—and most importantly, lower costs. The Mach-E and the Lightning were accepted to be money losers, mostly because of the audacious expense of their batteries. Ford needed to figure out how to extract massive costs from its next batch of EVs.

So there on one side of the room lay a pile of vehicle guts, the carcass of a Tesla Model 3 sedan. On the other side were the remains of the similarly sized Mach-E. Customers loved both. But Farley wanted to know how big an advantage Tesla had on its design costs—essentially the efficiency with which Tesla had engineered the car's power train. The question had nagged him for weeks, ever since he had pressed his engineering team on the point. He had been told there was no difference between the two.

Farley smelled bullshit. Tesla was selling about a million EVs a year and racking up profits, while Ford and other legacy automakers were bleeding losses on their electric cars. He had a strong suspicion that Tesla had a secret-sauce approach to engineering that resulted in a far better cost base to help offset the extreme expense of the batteries.

And now he had a guy who could help him find out. Doug Field had joined Ford a few months earlier from Apple, following a lengthy courtship by Farley. Field was a natural pick to lead this exercise: before Apple, he had spent five years at Tesla, where he was credited with development of the Model 3, the very car that lay cut open in the warehouse. It didn't take long for Field to identify the extra flab and waste tucked under the Ford Mach-E's taut body.

Weight is the enemy of electric-driving range, and shorter range hurts marketability. And the Mustang was carrying a lot of useless weight. It held an inefficient web of wiring harnesses to send electrical power throughout the vehicle. The Ford design added more than seventy pounds of unneeded weight to the car relative to the Tesla. The number of fasteners—bolts, screws, or rivets used to join components together—was 40 percent higher than the Model 3. Field pointed out brackets that were added to hold up parts that didn't need holding up, had Ford's team had put more fore-thought into the design.

For the Ford engineers in the room, the exercise was humiliating. The Mach-E had been revealed to the world at a splashy celebrity event in Los Angeles a couple of years earlier, and gushing media reviews helped boost sales. But strong sales don't always mean good business, and the SUV was far from profitable. Tesla had much bigger sales and was making money off them.

A big reason Ford was losing on every EV sale could be found in the tangle of vehicle entrails strewn across the concrete floor, staring back at Farley like a pool of red ink. In the end, the team discovered, Tesla held a $3,000–$4,000 cost advantage, simply by having a cleverer design for its internal components. A startup that hadn't existed when Farley got into the car business had out-engineered the company that invented the mass-produced automobile.

Farley turned and glared at his engineering chief, seething. The CEO felt like he had been lied to.[1]

. . .

Farley isn't alone. Just about every traditional automaker is going through its own messy reckoning with an EV revolution that presents huge opportunity and existential risk. The move to electric vehicles is the auto industry's biggest transformation since cars replaced horses early last century. Vehicles with internal-combustion engines emerged in the early 1900s as the mobility solution of choice for the world. Electric cars were an option then, too, which vied neck and neck with gas and diesel. The latter types won out in large part because Henry Ford's combustion-engine Model T took over the market—his design and manufacturing gave him the same kind of economic leg up that the Model Y seemed to have on the Mach-E. A global footprint of infrastructure sprang up around those fossil-fuel-propelled cars and trucks—highway gas stations, corner service centers, oil-lube shops. The technology was so dominant it hasn't been challenged much for a hundred years. It took more than a century, a climate crisis, and a rush of investment into green-energy projects for the first massive rethink of automobile technology.

The car business is perhaps the highest-profile sector in a swath of industries navigating the ripple effects from electrifying the economy and the broader response to climate change. Oil companies like Royal Dutch Shell and British Petroleum are investing in gargantuan batteries for energy storage. Developers are constructing buildings that rely on geothermal heat. Auctions for the rights to build wind farms off the Atlantic coast are fetching billions of dollars.

Few industries are up against change as fast and transformative as the one now scrambling the car business. Globally, drivers of cars, SUVS, vans, and light-duty trucks in 2022 pumped some 3.5 billion cubic tons of heat-trapping carbon dioxide into the atmosphere, about 10 percent of the world's greenhouse gas emissions.[2] But the car industry didn't come to this electric-car religion on its own. It had to be prodded and enticed.

Prodded by governments from Sacramento to Beijing and New Delhi, seeking to crack down on fossil fuels in an effort to stem the tide of climate change. In China, carmakers that don't sell enough plug-in vehicles can be barred from the market . The European Union has moved to ban gas- and diesel-burning cars starting in 2035. In the United States, the Environmental Protection Agency effectively mandated that more than half of new vehicles sold should be electrics by the early 2030s.

The regulatory tightening hasn't stopped at those major markets. One by one, individual countries, states, and even cities have also cracked down. South Korea declared plans to ban combustion-engine vehicles by the mid-2030s. Regulators in India want to outlaw diesel cars from big cities this decade. Want to drive your car on the Avenue des Champs-Élysées in Paris? By 2030, it had better be an electric—officials in the world's most visited city said diesel and petrol cars would no longer be welcomed inside the City of Light.[3]

But this transition is not just about governmental pressure. If that were the only factor, most car companies would simply be doing what they did for most of the last two decades: offer a few bland, uninspiring EVs and sell just enough to comply with the rules. Industry insiders literally called them "compliance cars." Think electric versions of the Chevy Spark, Fiat 500, or Ford Focus—small and plain, with electric ranges so skimpy that drivers would start sweating about running out of juice almost immediately.

A powerful force emerged, though, that eventually motivated just about every auto executive on the planet to bring their A games to the EV race.

Tesla rose from a few tech geeks strapping together laptop batteries in California to the ultimate industry disruptor. The company was formed in 2003 by entrepreneurs Martin Eberhard and Marc Tarpenning, who had sold an e-reader startup for $187 million and were looking for their next thing.[4] Months later entered Elon Musk, a South Africa–born inventor who was rich from sales of companies he'd had a hand in forming, including PayPal. He agreed to inject a much-needed $6.3 million into the cash-strapped car startup and eventually became CEO.[5]

Under the hard-charging Musk, Tesla survived multiple near-death experiences and then just kept coming. By the mid-2010s, affluent, younger, tech-savvy buyers—a demographic that carmakers drool over—were snapping up Tesla's sleek and speedy Model S sedan.

A startup automaker isn't even supposed to be a thing. The barriers to entry had been so massive—the supply chain and manufacturing so capital-intensive, and intricacies of the internal combustion engine so complex—that there hadn't been a new mass-market automaker in a century. But the emergence of electric vehicles shrank those hurdles. While batteries might be tricky and expensive, EVs ultimately are less complicated than combustion-engine cars. A small company can pretty much piece together some battery cells, an electric motor or two, and other off-the-shelf parts to build one.

Musk's company by 2019 would finally be mass-producing the $35,000 Model 3 sedan. The milestones piled up rapidly after that. It built a plant in Shanghai in a blazing-fast ten months, followed quickly by a factory in Berlin. The Model 3 would become the top-selling car in Europe, the birthplace of the automobile. A few years later, the Model Y SUV was on its way to becoming the number one car in the world.

Investors gobbled up all that good news. Tesla's valuation had already zoomed past that of Ford in April 2017. A week later, it topped GM's.[6] By the time of that Ford autopsy in early 2022, Tesla was the most valuable car company in the world. And not by a little. Its stock value topped $1 trillion, higher than that of Toyota, VW, GM, Ford, Jeep, Ram truck-maker Stellantis, and Honda—combined.

Car executives from Detroit to Tokyo had for years scoffed at that spiraling stock price. It made no sense. Tesla only sold a tiny fraction of the cars as the incumbents. For years it couldn't turn a profit. Musk was a loose cannon. To investors, none of that mattered. Year after year, VW, Toyota, and GM would crank out millions of cars and billions of dollars in profit. Relative to Tesla, their stock prices remained stuck in neutral. Wall Street had made its bet. The future was electric.

This book is not about Tesla. But it wouldn't exist without Tesla. Whether in name or not, Tesla's influence is on every page. The extent to which Musk forced traditional automakers down the electric path cannot be overstated. Musk has long insisted that Tesla exists to pave the way for sustainable, electric transportation by pressuring the rest of the industry to follow him.

"The car industry was operating on what I believe to be two false premises," Musk said at a conference in 2013. "One was that you could not make a compelling electric car—one that was aesthetically appealing, long range, high performance, all these things." And even if you did, "the car industry's opinion was that [consumers] will still not buy it."[7]

Tesla had disproved those, Musk said, so "we expect that other manufacturers will get into the electric-car market." That would take a couple more years. But after a few decades of head fakes and half-hearted attempts, the car companies finally went all in on EVs. By the early 2020s, the industry had outlined plans to spend more than $1 trillion through the decade on new EV models, factories, and battery plants.[8]

And that internal-combustion engine that had made the car business one of the biggest industries in the world? It was destined for a reasonably rapid fade into history. "I don't know where to spend money on them anymore," one GM executive said of internal combustion engines.[9]

You might be wondering what all the fuss is about. An electric car still has four doors and brakes and a dashboard. It's a simpler design mechanically and can be assembled more quickly than combustion-engine cars. Why would carmakers need to spend gobs of money and mental energy to figure this out? Why all the angst in that warehouse in Dearborn?

There are two big reasons why the transition to EVs is so difficult and fraught with risk for the legacy automakers.

First, it requires a massive rewiring of supply chains. Swapping out the guts of the car from an engine and transmission to battery and motors is like a combo heart-and-lungs transplant. And the stuff the automakers need for the transition to EVs is generally not the stuff that they are good at. Batteries, motors, and electronics—those are the domain of big Asian suppliers like LG and Panasonic and China's CATL. If the world gravitates to electric cars, GM's or Toyota's mastery of cylinder blocks and pistons and valve trains eventually won't matter.

Second, the batteries are still egregiously expensive, accounting for as much as one-third of an EV's cost. Claims about some imminent breakthrough that will slash battery expenses never seem to pan out; getting costs down likely will continue to be an incremental grind over years, just as getting better miles per gallon out of a combustion engine has been a decades-long grind. That puts tremendous pressure on the engineering staff of a car company to find more-efficient ways of stitching together that giant rolling battery called an EV. As of the mid-2020s, the state of play remains bleak for most of the industry: Tesla, a few Chinese automakers and a few luxury brands are still the only ones turning a comfortable profit on EVs.

For a while, it looked like this transition was going to hit escape velocity. Traditional carmakers for the first time began offering consumers electric cars that were fun, cool, and fast. New models like Ford's Lightning and GM's Cadillac Lyriq SUV were racking up multiyear wait lists. New entrants like startup truck maker Rivian also were hot. Buyers who were lucky enough to be next in line for an EV often forked over $10,000 to $20,000 above the sticker price. Investors were falling over themselves to bet on a coming EV boom, bidding up shares of even the flimsiest startups.

EVs surged to 18 percent of all new vehicles sold globally in 2023, up from about 2 percent just five years earlier.[10]

This frenzy, perhaps predictably, wasn't sustainable. By early 2023, some of those reservation holders lost interest and faded away. Musk began slashing prices to stoke demand, leading to a global price war and deeper red ink for nearly everyone. Car dealers began to lose interest. Global EV sales still were growing, but at a much slower rate. The tech bros and other early adopters already had their EVs. The industry needed to appeal to a more discerning buyer, one who wasn't ready to pay 20 grand above the sticker price or stop for an hour to charge the car on the next road trip. "EV euphoria is dead," CNBC said flatly in March 2024.[11]

That all sounds grim. So what's with the title of this book? How is this EV transition inevitable?

Let's get this out of the way: combustion-engine cars are going to be around for a long time. Twenty years from now, an American car shopper probably will still be able to walk into a showroom and buy a car with a gas engine—there may even still be ample selection. Even though surveys show that around half of US buyers are open to an EV, there are still too many places where they will be a tough sell. There are rural and suburban areas where chargers are few and far between, and people drive longer distances. Pickup truck owners who tow boats or horse trailers up mountain passes probably won't be among the early adopters. Apartment dwellers are at a big disadvantage because 80 percent of charging is done at home.

Skeptics have a laundry list of reasons why the EV transition is doomed. There aren't enough public chargers—especially in the United States—and won't be for a long time. The ones that do exist are often broken, can't connect, or don't sync with the app. Some experts say the industry won't be able to pull enough minerals from the ground to produce the batteries needed for EVs to reach mass production. The electric grid is another concern—how will it be able to handle millions of EVs sucking juice every night, at a time when a boom in artificial intelligence has huge power demands of its own?

Then there are questions about whether EV technology is really ready to be anything more than a niche market for tech-forward buyers. Batteries don't like very cold weather, which reduces range and can disrupt the charging process. Media coverage of Teslas with bricked batteries in subzero Chicago in the winter of 2024 became a talking point for conservative media for weeks. There is also the fire risk: in rare instances, EVs combust seemingly out of nowhere. The questions continue straight through to the end of an EV's life cycle: What will we do with all those spent batteries? Isn't that its own environmental quagmire waiting to happen?

All of those factors pose real obstacles to the EV revolution. It seems clear now that the big, provocative predictions made about the market going all electric by 2035, or how some brands would sell only electrics by decade end, didn't age well. Hardly any industry insiders now think gas cars will be completely phased out in favor of electrics within a decade.

But a funny thing happened as the red-hot market for EVs began to cool: demand for hybrid gas-electric cars took off. It turns out a lot of those EV-curious car buyers weren't quite ready to make the leap to a fully electric car. But they were open to the idea of electrons helping to power their way to the office or their kid's soccer practice or on a family road trip. Many of those hybrids are the plug-in variety, able to travel in full electric mode for a few dozen miles.

What's that say about where EVs are headed? For a good road map, look back twenty years, when a Toyota Prius driver had to have thick skin. Toyota's fuel-sipping hybrid with revolutionary efficiency was considered a nerd-mobile by many—only for tree huggers or virtue-signaling celebrities. Today, nobody blinks at someone who drives a hybrid. It's just another option in the showroom, one that looks just as nice as a fully gas-powered car but happens to get 40 percent better fuel economy. Fast-forward a decade or two—as more people have more friends and family members who own cars with plugs—and the same normalizing effect will happen for EVs.

It's also instructive to look at the experience of Tesla, which is the world's most fully evolved version of an electric car. Go chat with Tesla owners and ask them what they'll buy for their next car. Nearly nine in ten say another Tesla.[12] Do you think the other 20 percent are going back to gas? It's unlikely.

"After a few miles in an EV," *Wall Street Journal* car critic Dan Neil wrote in 2024, "going back to internal combustion feels like returning to whale-oil lamps."[13]

The superior efficiency of EVs is hard to ignore in this debate, too. In a gas-powered car, the majority of the fuel that is put in—as much as 75 percent—ends up wasted, primarily used up as heat inside the engine. In electric cars, energy losses run only 15 to 20 percent.

But sure, for the American car shopper, that combustion-engine SUV or pickup truck will probably be there in the showroom a decade or two from now. While US consumers generally are intrigued by EVs, it's clear that the frenzy of the early 2020s has died down, and most are fine waiting for the messiness to get sorted out. And some remain flat-out hostile toward EVs, viewing them through a political or ideological lens.

This state of play leaves GM, Volkswagen, Toyota, and the other global car titans with a choice. Do they soothe themselves with the fact that EV sales have hit a lull in their corner of the world? Does that mean they can scale back on the tens of billions they've already directed to the EV switch—money poured into new supply chains, new designs, new factories? Will they forgo massive incentives created by the government to get going on EVs? Will they stop working hard to get battery costs down, or make batteries last longer? Will they stall, again, on developing an affordable EV for the masses?

Ford's Farley was confronted with these questions on a conference call with Wall Street analysts in February 2024. Investors were growing worried about all this capital flowing into EVs when it wasn't clear if enough consumers were ready. Maybe the company should consider pulling back and steering more investment to big hulking pickup trucks for business customers. "Almost all the profits are funding this EV science project," one analyst groused.[14]

Of course, Farley knew doing that would help Ford's bottom line and probably its stock price in the short term. But, as a student of automotive history, he also knew what happened the last time American carmakers ignored shifting market dynamics. Japanese automakers in the 1970s and '80s blitzed the US market with smaller, fuel-efficient imports, and eventually set up shop with American factories. The Detroit automakers barely survived and have never regained that lost market share. This time it's China. No matter what Farley and his peers decide, Chinese companies are not scaling back. With plenty of financial backing from the Chinese government, they are building new supply chains and new factories. They are developing better, cheaper batteries. The Chinese government is footing the bill on charging infrastructure. In 2023, EVs and plug-in hybrids—which operate in electric mode for a few dozen miles before switching to gas—approached 40% of new vehicle sales in China, compared with around 21 percent in Europe, 10 percent in the United States.[15] It's not a coincidence that the place where the pain points of EV ownership—high prices and charging headaches—have been largely removed, consumers have embraced them. The most prominent Chinese maker is BYD, and it's already about as big as Tesla. Cargo ships laden with low-cost, attractive and tech-laden Chinese EVs have landed in Europe. There is talk of factories in North America.

For the first time ever, what legacy carmakers decide to do isn't necessarily what the market will do. The threat from Chinese EV makers and the many other newcomers is greater than any previous threat to these companies. It's existential.

"The EV customers are very loyal and they love the vehicles," Farley told the investors. "So it's on us to get the cost right. That is the issue with the transition."

He added: "The journey on EVs is inevitable in our eyes."[16]

1

GM's Last Stand

Under a gray Michigan sky, a pickup truck sat like a gleaming white jewel amid a drab landscape of cargo trailers and loading docks outside General Motors's massive Factory Zero assembly plant in Detroit. The prototype electric GMC Hummer looked like it had been dropped from a futuristic battlefield. The broad-shouldered truck suddenly sprang forward from a standstill, chalky dust and smoke billowing up from under its massive tires. The pickup screeched to sixty miles per hour in a blink, before lurching to a stop. A few seconds later, the driver's window lowered to reveal a wide-eyed, white-haired, seventy-eight-year-old man in the driver's seat, dressed in a trim navy blue suit and maroon tie.

"This sucker's something else!" President Biden gushed to a handful of TV cameras and reporters. In the back seat, GM CEO Mary Barra sat beaming. Having the president of the United States gush about your product isn't a bad way to introduce the vehicle that would loudly announce the biggest strategic pivot in your company's history.[1]

It was November 2021, a month before GM was to begin rolling out the Hummer as the first in a lineup of new EVs. GM's team wasn't counting on it to be a big seller. With a $113,000 starting price, the pickup was destined to be a toy for the rich. And yet, the hulking truck was critical to GM's electric-vehicle story. Barra saw the Hummer as a statement, an EV that would stoke real excitement. She was painfully aware of the failings of GM's past EVs. If there was a theme to those efforts—names like EV1, Volt, and Bolt—it was that they were, well, a bit dumpy. While Elon Musk's Tesla was coming out with sexy roadster convertibles and sleek, stylish sedans, GM was putting forth the automotive equivalent of nurse's shoes.

With the Hummer, the GM team went in the opposite direction—it may have even overcorrected. Its one-thousand horsepower was three times that of a typical big pickup. At around nine thousand pounds, it weighed nearly three times more than a Corvette. The president's blurring sprint that day was thanks to the truck's Watts to Freedom mode—the cheeky WTF acronym a play on the mind-blowing thrust of such a massive truck hurtling itself to sixty miles per hour in about three seconds, as fast as a Ferrari.

GM engineers slapped a kitchen's sink worth of features onto the Hummer. They installed a rare four-wheel-steering system that enabled the rear wheels to turn in the same direction as the front two, allowing the truck to essentially slide diagonally. Ostensibly the feature was to help drivers park the beast more easily and avoid boulders or other tight spots on the trail. In reality, it mostly made for cool marketing. "Crabwalk mode" was featured in a Super Bowl commercial showing crabs scuttling across the sand diagonally, set to the catchy song "Get Ur Freak On" by Missy Elliott.

Applying the Hummer badge, though, might have been a more in-your-face move than any of those engineering feats. Fifteen years earlier, the gas-guzzling, military-inspired Hummer had become a symbol of American excess, lubricated by years of cheap fuel and heady stock-market gains. In 2010, amid the somber aftermath of GM's bankruptcy and the nation's financial crisis, the Hummer was quietly phased out.

Biden's Factory Zero visit was intended to showcase the dawning EV factory boom in America. The president had just pushed through one of his signature measures, a $1 trillion infrastructure spending bill that earmarked $7.5 billion to string EV charging stations through lonely stretches of the American heartland. But he was still trying to rally support for an even larger spending package that called for pumping up the $7,500 consumer EV tax credit to as much as $12,500—only, though, for

American-made electrics built by union workers, an unabashed play for the labor-union vote in the 2022 midterms.[2]

Inside the sprawling plant, the mood was festive. Hundreds of GM factory and office workers sat side by side with elected officials in white folding chairs under bright lights on the factory's polished concrete floor. Ceremonial patriotic bunting draped the stage. Large blue-and-red banners proclaimed, "A Future Made in America," along with an image of the Hummer's bawdy front end. A Detroit high school band blared Motown hits.

Moments after Biden floored the Hummer into WTF mode out back, he stood inside the factory amid a loose circle of GM executives and plant workers. Before them was the husk of a white, work-in-progress Hummer, hoisted onto an orange steel lift. A tangle of wires grew from its hood and underside. One UAW worker explained how the team tests the truck's electrical current. Another explained something about a data-communications link. The president listened intently, even though the impromptu technical briefing would have been lost even on an engineering grad.

At one point the plant's executive director, Jim Quick, cut in: "We'll be shipping units to customers by the end of the year," he told the president through a light-blue face mask, as Covid-19 was still lingering in public gatherings at the time. "We were able to do that with immense speed, faster than anyone is able to do in the industry." That statement, though, wouldn't age well.[3]

At this point, Biden, a self-proclaimed car guy, got reminiscent. He grew up hanging out in the car dealerships his father managed. He recalled going down to the auto auctions, where used cars were bought and sold at a stock-trader's pace. Returning to the real purpose of the visit—appealing to union labor—Biden rattled off his home state of Delaware's auto-industry bona fides. GM's Wilmington, Delaware, assembly plant was once home to thousands of UAW workers, churning out small cars for the Chevy and Saturn brands before closing in 2009.[4] "We don't have any of it now," he said. The UAW, he went on to say, was "the first outfit to ever endorse me."

The president wanted this visit to show that his economic and climate policies were helping faded old factories spring back to life with fresh EV production. But for both Biden and Barra, it was a complicated message. Broadly, the industry's EV transition was expected to cost jobs, union officials fretted, potentially tens of thousands of them, according to a union-funded study. About two years before Biden's appearance, the factory was dark while picketing workers warmed themselves at barrel fires during a strike that was, in many ways, about protecting jobs in the EV transition.

As Stevie Wonder's "Higher Ground" pumped through the loudspeakers, rattling the factory's steel innards, Biden strode alone to the blue podium on stage. The crowd clapped and whistled. "Hello Detroit!" he yelled to cheers. "I never could have imagined vehicles like the one I just took for a spin. Proof that America has what it takes to win the competition of the twenty-first century!"

· · ·

GM, in many ways, invented the modern auto industry. Bitter rival Ford had the clear lead in the early twentieth century with its Model T and proliferation of the assembly line. But GM was never far behind and, by the late 1920s, passed Ford as the world's largest carmaker by sales.[5] GM introduced the electric starter, the pollution-reducing catalytic converter, the airbag. On the marketing side, it came up with annual model updates while offering a spectrum of brands and choices suited for different tastes, as Ford stuck to a one-size-fits-all approach with its black-only Models T and A.

GM's identity was built on sports cars like the Corvette and hulking trucks like the Chevy Suburban. At times, GM's engineering prowess combines with its massive scale to achieve something special. Take the eighth-generation Corvette, released in 2020, which, for the first time in the car's seventy-year history, moved the engine from under the hood to the car's midsection for more-precise handling. The move put the value-conscious sports car in a class with Porsche or Ferrari at a quarter of the price.

Occasionally, GM applied that technical muscle to electrification. In 1987, an offbeat group of California geniuses built a solar-powered car that allowed GM to beat twenty-two other competitors in a highly publicized, 1,867-mile race across Australia, reaping praise as an engineering leader.

GM teamed with aeronautical engineer and inventor Paul MacCready to develop the Sunraycer, a round-nosed, space-capsule-like vehicle built using an ultra-lightweight aluminum chassis and Kevlar body. GM's research arm developed the motor, and its Hughes subsidiary supplied the silver-zinc battery to store the solar power. The experimental machine cruised across the Australian Outback in 44.9 hours at an average speed of 41.6 miles per hour, blowing away the competition.[6]

As an encore, the team proposed an all-electric car, and GM's management agreed to back the project. The result was the EV1, a space-age-yet-dorky-looking two-seater with lead-acid batteries stuffed behind its two seats, allowing the car to travel up to a hundred miles on a single charge. Practically, it was the invention of the modern mass-produced EV.

The car delighted environmentalists and California's regulators, which had passed a law that would eventually require automakers to sell electric cars in the Golden State. But GM fell short of the mass-produced part: it made only about 1,100 EV1s from 1996 to 1999, far below the company's projections.[7]

In what some now see as an epic misstep, GM walked away from the project, deciding it was too big a financial drain. It was the start of a will-they-or-won't-they dalliance with EVs that has been maddening for green-car advocates. Just when GM seemed to advance down the path to an electric future, it would squander the opportunity and cede its lead to others. To Toyota. To Tesla. To startups like Rivian and Lucid. Even, in the early part of this latest EV race, to archenemy Ford.

GM's on-again off-again relationship with EVs stems from the EV1 experience, which wasn't just an expensive failure. It was a public-relations disaster.

Around the year 2000, GM began refusing to renew leases for hundreds of EV1 customers. In the ensuing years, the company took the cars back one by one, often over the fierce protests of EV1 owners who begged to buy out their leases. But GM cited liability concerns and issues with continuing to service a car made in such small numbers.

GM had managed to turn a groundswell of goodwill into a corporate-image black eye as it began crushing the EV1s in Arizona. Owners, including *Baywatch* actress Alexandra Paul, held vigils in protest.[8] Paul was even arrested after trying to block a truck taking EV1s to be destroyed. Some owners held a funeral for their EV1s, driving the cars in a procession around the grounds of a Hollywood cemetery to the piercing wail of bagpipes.[9] The EV1 debacle spawned an unflattering documentary, "Who Killed the Electric Car?" and made GM a relative villain of the environmental movement. For years afterward, talk of electric vehicles, never mind invention or production of them, was verboten inside GM.

That is, until Maximum Bob showed up.

. . .

Tall, silver-haired, seventy-four-year-old Bob Lutz spent a crisp California morning in October 2006 tooling around the beaches of Camp Pendleton marine base, north of San Diego, in a zippy minivan. But this wasn't any van. It was the Chevy Sequel, a GM prototype that was fueled solely by hydrogen fuel cells, which smash together hydrogen and oxygen to produce electricity to propel the wheels, emitting no pollutants—just harmless water.

The Sequel was a blast to drive. The instant torque rocketed the van forward with a tap of the accelerator in a silent thrust as Lutz and a team of engineers, and automotive journalists put the prototype vans through their paces. At one point some Marine officers invited Lutz to hop in one of their amphibious assault vehicles and dart into and out of the surf.

The occasion should have been nirvana for Lutz, who five years earlier, in the twilight of a remarkable career in the car business, was tapped as GM's global product chief and tasked with nothing short of resuscitating GM's ability to make great cars, which had eroded through the time of the EV1 saga and into the early aughts. Here, he was in his element, with an opportunity to drive machines with reckless abandon in the presence of his beloved marine corps, which decades earlier had taken him as the troublemaking rich-kid son of a Swiss banker and turned him into a steely-nerved fighter pilot. And, as GM's global product chief, it was a chance for the famously verbose executive to gab with the press.

But even as he smiled while zipping around in the hydrogen-powered van that morning, Lutz was a reluctant participant.[10] Around the time, he had been waging an internal battle against hydrogen fuel cells as the focal point of GM's future green-car strategy.

Though ancient by auto-executive standards, Lutz was no humbled graybeard content to fade into an overdue retirement. On weekends, he once explained, he still flew his collection of decommissioned fighter jets sourced from Czechoslovakia and Germany out of executive airports around Detroit "to bore holes into the sky."[11] He sometimes commuted to work in Detroit the 45 miles from his home near Ann Arbor by flying his own helicopter. He earned the nickname Maximum Bob.

Lutz was neither an engineer nor vehicle designer. But he was legendary in the car business for developing breakthrough models that revitalized brands. At BMW, in the 1970s, he was instrumental in elevating the German car company's 3 series nameplate into a global luxury juggernaut. In the eighties, at Ford, he initiated the development of the Explorer, which helped make SUVs into America's new family vehicle.

In early 2001, after his retirement from Chrysler, Lutz's arrival at GM was unexpected. At the Detroit auto show that year, Lutz quipped that some of GM's recent models looked like "angry kitchen appliances."[12] Soon after, then-GM CEO Rick Wagoner hired him. Wagoner was desperate to reverse decades of sliding market share amid criticism of bland vehicle designs and mediocre quality.

Lutz arrived at GM to find nothing maximum about it. It was a stultifying culture in vehicle development that prioritized process and pragmatism over emotional design and performance. Most of the yet-to-be-revealed new models he reviewed he described as a "horror show."

"The system created research-driven, focus-group-guided, customer-optimized transportation devices," Lutz wrote in a 2011 book.[13] "Designers were now reduced to the equivalent of choosing the font for the list of ingredients on a tube of Crest."

Lutz quickly injected some flair. Among his first projects was a big Cadillac prototype sedan equipped with an absurdly large 16-cylinder engine and silk floor mats. The Cadillac Sixteen was only a concept car—it never went into production—but it served to show that GM could still bring it. With time, GM began winning some accolades from the automotive press, which tends to gush over striking designs—murky business case be damned.

Lutz was a warrior, but not an eco-minded one. A staunch conservative, he was quoted as having called the concept of manmade global warming a "total crock of shit."[14] But his green-car passions were fueled by two things: his deep contempt for an old rival, and his curiosity about a new one.

Around the time of the Camp Pendleton drive, Toyota's Prius hybrid had emerged as an unlikely smash hit. The fuel-sipping hatchback, which combined a battery pack and electric motor with a small engine to achieve eye-popping fuel-economy numbers, hit the market just as the Iraq War jacked up prices at the pump. It became a status symbol in car-conscious, environmental-leaning Los Angeles, influenced by celebrity owners such as Leonardo DiCaprio and George Clooney.

For Lutz, the Prius represented the glowing reputation that Toyota had come to enjoy at GM's expense. The Japanese automaker had been devouring US market share for a few decades by then, with efficient cars that garnered a sterling reputation for quality and reliability. Now, Toyota was emerging as an environmental champion, too. It was all too much for Lutz.

"Toyota was not only selling hybrids, but more importantly, they were gaining this valuable reputation as a technologically advanced and environmentally virtuous company that could do no wrong," Lutz would recall later. "Their cars were better, their gas mileage was better, their technology was better, their executives were paid less than ours. Everything about Toyota was just divine. And I was getting sick and tired of it."[15]

Meanwhile, Lutz's sense of urgency on battery-powered cars was coming from a Silicon Valley startup that had been generating buzz in the automotive press: Tesla. The company, then led by CEO Martin Eberhard, had a few months earlier unveiled its Roadster, a small two-door sports car—essentially the frame of a Lotus, festooned with thousands of lithium-ion laptop batteries, strapped together into a single pack. As a smattering of car magazines had been writing with glee, the Roadster had so much instant torque off the line, it could outrun a Ferrari to sixty miles per hour. Nothing got Lutz's attention more than head-snapping speed—and rave reviews from car magazines.

Lutz was convinced that electric cars were the optimal green-car solution of the future. Hydrogen-fuel-cell technology, while holding long-term potential for the trucking industry, was a money pit of false promises, he thought. Of course, Lutz wasn't at GM during the development of the EV1 program, a debacle that drained $1 billion from GM's coffers.

There's a lot to like about hydrogen-fuel-cell vehicles, which hold the greatest promise for long-haul trucking. Hydrogen gas is stored in on-board tanks and can create enough electricity to power a vehicle for long distances. Importantly, refilling can be done in a few minutes, similar to gasoline, so owners would be less intimidated by that than having to adapt to a foreign procedure, like EV charging. The problem is, good luck finding a place to fill up—the cost of refueling stations can run into the millions, and few exist outside of California.

GM had a deep history and experience in fuel cells. As far back as the 1960s, GM had developed a prototype show car called the Electrovan. Engineers stuffed thirty-two fuel cells, connected with 550 feet of plastic piping, into a GMC Handivan, which later became an inspiration for the Mystery Machine van driven by the motley crew in the animated *Scooby Doo* series. The Electrovan was revealed in 1966 but soon scrapped.[16] Decades later, GM had logged zero commercial sales of hydrogen-fuel-cell cars.

During the lunch break at Camp Pendleton, Lutz held court with reporters, dutifully spouting GM's ambitious plans to commercialize hydrogen fuel cells. They would be in showrooms within five years, he declared, and had the potential to revitalize GM's innovative spirit and reputation with American car buyers. Later he sidled up to some of GM's hydrogen-fuel-cell specialists and electric-vehicle experts—out of earshot from the journalists—to talk about what was really on his mind: batteries.

"Please explain to me again why you think lithium-ion batteries aren't the answer for automotive applications?" Lutz asked. His query was greeted with engineering rationalizations for why the energy source—just then taking hold in laptops and cell phones—wasn't viable for cars. It was too volatile and susceptible to catching fire, he was told. The batteries weren't capable of generating enough torque—abrupt acceleration from a standstill—and weren't suitable for sustaining higher speeds efficiently.

Lutz returned to Detroit, unconvinced. A couple months later, he got wind that Eberhard was in town for the Detroit auto show, which was happening down the street from GM's headquarters. Lutz dispatched someone from his staff to run over to the show to track Eberhard down and invite him for an informal meet-and-greet. At the time, Eberhard, had his hands full. The engineer was trying to launch the Roadster under tremendous pressure from a huge order backlog and heat from Elon Musk, then Tesla's biggest investor and a director. Still, Eberhard agreed to swing by the Renaissance Center, GM's silver, cylindrical-towered headquarters, the iconic skyscraper on the city's skyline. He met with Lutz and high-ranking GM engineer Jon Lauckner in Lutz's office for a few hours.

Eberhard made clear he wouldn't be divulging trade secrets but was up front about Tesla's experience with lithium ion to that point. "Isn't it too volatile for cars?" Lutz asked. Eberhard conceded that there had been cases of "spontaneous dis-assemblies," a euphemism for battery fires. But the Tesla team felt like they had managed to mitigate the risks. And, based on the Roadster's lightning-quick acceleration, it was clear that lithium ion could deliver plenty of performance.

Lutz and Lauckner huddled again in Lutz's office within weeks of the Eberhard meeting. Clutching an expensive, old-fashioned fountain pen, which he had no idea how to use, Lauckner splattered ink as he feverishly sketched his version of how a future so-called plug-in hybrid could work. A lithium-ion battery pack would provide around forty miles of pure-electric driving range. Once that was depleted, an onboard, gas-powered motor generator would kick on to charge the battery and deliver power to the electric motors, providing another three hundred miles or so of normal driving range. Lutz watched the excitable Lauckner's knee gyrate up and down as he scribbled across the page. Within fifteen minutes, the concept for the Chevrolet Volt was born.

Lutz took it to CEO Wagoner, who was dismissive. "Bob, we lost $1 billion on the EV1. Are you seriously proposing we get back into electric cars?" Wagoner asked.

"C'mon, Rick, you know me. I'm not going to risk losing billions of dollars. This is one project," answered Lutz, who, even with his gun-slinging, iconoclastic style, had a reputation of being careful with capital budgets. "I'm proposing we build one show car as a proof of concept, to draw a line in the sand and say, 'Hey, if other people can do hybrids, we could do a much better hybrid. If other people can use lithium-ion batteries, we know how to do that even better.'"

Wagoner relented and allowed Lutz to pursue Lauckner's sloppy-ink creation. The way Lutz saw it, GM couldn't just do a slightly better version of the Prius. He could already predict the headlines: "GM Launches Hybrid to Counter Prius." Lutz concluded that GM needed to wow people. He has even compared his competition with other carmakers to the space race. "When the Soviets put a dog into orbit, we should have countered by putting a guy into orbit," he said. "You don't gain points by matching your opponent. You need to clearly outdo your opponent."

Getting Lauckner's concoction to market would indeed be a moonshot. At the time, hybrid cars deployed relatively small batteries to assist the gas engine and reduce fuel consumption—no major car company offered a hybrid for the US market that allowed the vehicle to be driven on electric power alone. The concept of an electric car that could keep going even after the battery was depleted was novel. GM eventually came to call it an "extended-range EV." But in fact, it was vastly more complex than an EV because it essentially required two power trains—the combination of parts and software that propel the vehicle—instead of one: the EV part, with large battery pack, electric motors, and wiring harnesses, plus a gas engine, a complex web of planetary gears, and a bunch of computer code to allow the two to communicate.

Lutz assembled a team of about two dozen designers, engineers, and marketing people to launch the project, initially called the iCar, as a nod to Apple's then-revolutionary iPod. He appointed veteran GM engineer Tony Posawatz to run it. A GM lifer at the time, Posawatz spent much of his career working on GM's gas-guzzling trucks and SUVs, including its brawny Cadillac Escalade, among the least-green cars on the planet. The Dartmouth MBA had an unusually effusive personality by engineers' standards and an entrepreneurial bent. Lutz figured Posawatz would be a natural cheerleader to help the team persevere through the inevitable internal GM resistance that Lutz rightly anticipated.

The group wasn't quite a secret but was treated like one. In the early months of development, the team met in a small, out-of-the-way conference room deep inside a design building at GM's massive Tech Center campus in Warren, a twenty-minute drive north from GM headquarters.

In a lightning-fast eight months, GM had a show car prepped to debut at the 2007 Detroit auto show. Even though it had four doors, the car had a two-door Roadster look to it. It looked lean and fast, with ample sheet-metal creases, prodigious haunches, a long, lean nose, and a canopy-like glass roof.

"What you see before you is a truly futuristic concept: an electric vehicle that uses a gas engine to create additional power," the BBC's TopGear said in a review. "The shocking bit? It's American."[17]

Normally, it takes about five years for a brand-new car model to go from an idea to the showroom. Lutz—still seething about the Prius and anxious over Tesla's ambitions—wanted it done by 2010. To do that, though, there would have to be sacrifices to that head-turning design.

For starters, the exterior sheet metal was going to require massive changes from the original, sleek silhouette. Although the show car looked slender, engineers were shocked when they put it in the wind tunnel for the first time to test its aerodynamics, a critical consideration for an electric car to conserve battery range. The aero measurements for the taut, muscular concept car were so bad, "we thought maybe we put it in backwards," Lutz quipped.

GM executives decided that, both for aerodynamics and cost reasons, the car would have to be built on the same platform, or basic frame, used by some of GM's high-volume small-car models, including the Chevy Cruze. That meant big changes. The long, lean front end was replaced with a rounded snub nose. That vast panoramic glass roof wasn't going to work.

The end result was a pedestrian-looking hatchback, not unlike one of the dozen compact gas-powered cars on the market at the time—and not a whole lot different from the Prius. It was a far cry from the sultry design that Lutz had championed. GM's product-development machine had shoehorned it into the engineering-centric system that would maximize cost savings and efficiencies. And that system was one developed and refined over a century to make gas-powered cars. Beyond Lutz's considerable influence, there wasn't enough industrial will inside the vast GM enterprise to build an electric car from scratch, without internal-combustion-engine DNA.

Lutz and Posawatz knew the eye-catching car they envisioned was slipping away. But they persevered in their roles as quarterback and cheerleader of the project.

On September 16, 2008—GM's one-hundredth anniversary—the production version of the Volt was unwrapped during a boisterous party inside the Renaissance Center's Wintergarden, a five-story atrium of glass windows offering expansive views up and down the Detroit River and across to Windsor, Canada. Before the unveiling, hundreds of journalists and GM employees, many in dark business suits, sat through an hour-long GM business update.[18] The slick PowerPoint presentation avoided mention of the company's darkening financial picture. This was the day after Lehman Brothers had collapsed.

GM had lost an eye-popping $18 billion in the first two quarters of the year, heaping pressure on Wagoner to deliver signs of a long-awaited turnaround. By that time, GM lobbyists in Washington, DC, had been whispering into the ears of Michigan's congressional delegation about the idea of a potential federal bailout to keep the automaker from tipping into bankruptcy. At the event, Wagoner relished the reprieve from the drumbeat of ominous news about GM's bleak outlook. Unveiling a sultry new model was just the balm the company needed.

Except, as the huge digital screen that Wagoner had been presenting on parted in two, the silver Volt that rolled onto the stage—piloted by Lutz behind the wheel—looked unrecognizable from the original concept. This version's bulbous design, with piano-key-black trim outlining the windows and a silver, metallic pattern for the grille, wasn't objectionable or ugly. It was just . . . plain. "The final result is considerably more conventional" than the flashy show car, veteran auto writer Jim Motavalli wrote in the *New York Times*. "Think of a more streamlined Saturn Aura or Nissan Altima."[19] For Lutz, Motavalli's pen might as well have been a dagger.

Still, most of the media coverage from the day was positive, focused more on the Volt's groundbreaking power train than its design. GM executives gushed about the Volt being the company's "moonshot." The sixteen-kilowatt-hour lithium-ion battery would deliver about forty miles of electric range, before a small gas-powered generator kicked on to deliver electricity to propel the wheels. GM executives pitched it as the perfect commuter car. With the average one-way commute of American workers less than thirty miles, most workers should be able to get to work and back purely on battery power, or less than a buck a day, without ever having to stop at a gas station, they said.

MotorTrend called it GM's "great green hope."[20] CNBC noted that the Volt would be the first electric car to hit showrooms from a mass-market brand. "If the Volt is first AND works," the network's auto reporter Phil LeBeau wrote, "GM could capture the halo effect of leading the electric car revolution."[21]

The Volt's high-tech setup would still be in development for another two years. It would be a traumatic time for GM, and a time that humbled the notion of infallibility for American blue-chip corporations. Six months before the Volt launch, in March 2009, Wagoner was gone, pushed out by the Obama administration task force set up to handle the bailouts of GM and Chrysler. In June that year, GM, once the world's largest and most profitable corporation, filed for bankruptcy. Longtime GM employees lost all their stock. Dealers' franchises were cast into uncertainty.

Inside GM, vehicle programs that were under development were being gutted amid thousands of layoffs, as GM sought to save cash. The gestation period for one new car model can take more than five years and involve hundreds of employees from design, engineering, and purchasing. And yet, the Volt project—one that even GM executives conceded was sure to be a money loser—was untouched. The team was racing to get the Volt into showrooms in the second half of 2010, before Nissan rolled out its Leaf, to win the bragging rights to be the first affordable EV to market.

To build the Volt, GM poured more than $300 million into its Detroit-Hamtramck assembly plant, which, more than a decade later, would be renamed Factory Zero, a play on the zero-emission EVs to be built there, including the Hummer. The plant straddles Detroit and the densely populated city of Hamtramck, a historically Polish enclave that saw an influx of

Arab Americans since the 1990s and teems with immigrants from Yemen and Bangladesh. The big, barge-like sedans that had been built there for decades—the Pontiac Bonneville and Cadillac DTS—had fallen out of favor and were phased out. By the time the Volt came along, the Detroit-Hamtramck assembly plant was barely operating.

The car got rave reviews when it finally arrived at dealerships in late 2010. It wasn't going to solve GM's financial problems or remove the stain of bankruptcy. But automotive and tech journalists marveled at its unique ability to zip around town without using a drop of gas, while still being capable of leaving town without range anxiety.

"It's a dramatic reinvention of the great American car, without sacrificing the great American road trip," *Popular Mechanics* wrote.[22]

"A bunch of Midwestern engineers in bad haircuts and cheap wristwatches just out-engineered every other car company on the planet," *Wall Street Journal* columnist Dan Neil wrote.[23]

President Obama was so intrigued by the car that he took it for a spin on the White House grounds, breaking the Secret Service's ban on presidents driving.[24]

At the same time, the Volt became a symbol of Obama's push for electric cars and a lightning rod for critics of GM's government bailout. A $7,500 tax credit implemented under his administration made the Volt a target of Republican attacks. In November 2010, *Washington Post* columnist George Will wrote of the Volt: "People will have to be bribed, with other people's money, to buy this."[25]

Republican Newt Gingrich, in his stump speeches for the 2012 presidential nomination, told crowds, "You can't fit a gun rack in a Chevy Volt." After an owner uploaded a video to YouTube showing he could, in fact, fit a gun rack in his Volt, Gingrich issued something of a correction. "Yeah, you can get the gun rack in the trunk," Gingrich said. "But then where do you put the deer?"[26]

Lutz had accomplished nearly everything he had set out to do with the Volt. GM was lavished with praise for its technological accomplishment. It helped move the storyline forward from GM's ugly federal bailout, if only just a bit. GM had out-Toyota'ed Toyota.

But deep down, Maximum Bob knew he had come up just short on the one element that he has always put above all else on vehicle programs: an emotional, passionate design. Lutz's instincts had been right to push the GM machine hard to make a head-turning car that would have the gravitational pull of a device like an iPod. The Volt needed to be a car not just for tree huggers and engineering nerds, but for Silicon Valley venture capitalists, Manhattan bankers, and attorneys in the heartland—people who weren't just interested in saving the planet but looking cool doing it.

The Volt and Nissan's Leaf both rolled out toward the end of 2010. By then, though, Tesla had already begun taking orders for an electric car with a sinuous silhouette—one that Lutz would surely have salivated over. Within two years, car reviewers at the influential *Consumer Reports* magazine would call Tesla's Model S the best vehicle they ever tested.[27]

The Volt would go on to garner a cult following, but not a whole lot else. Through the 2010s, as Tesla steadily took off, the Volt couldn't overcome its boringness, and sales fell short of GM's forecasts. In 2016, the company released a fully electric car, the Chevy Bolt, envisioned as a rival to Tesla's breakthrough Model 3 sedan, which came out soon after. It didn't take long for GM executives to figure out that wouldn't be much of a race—the Tesla would soon be outselling the Bolt by ten to one. Mary Barra, by then GM's CEO, knew GM needed to get serious.

. . .

Barra was an unlikely figure to set in motion GM's biggest strategic pivot in its history. She was, after all, a company lifer, having started as an 18-year-old co-op student, inspecting fender panels at a Pontiac factory. Barra grew up immersed in a corporate culture known for its bureaucracy and inaction.

An engineer, Barra spent much of her career inside GM's clattering factories, far removed from the fun stuff like design studios, test tracks and glitzy auto-show reveals. In the late 1990s, following a bruising, 54-day strike at a GM factory in Flint, she was dispatched to smooth things over with UAW officials. Her bosses figured Barra's factory background and even keel gave her credibility with the union.

With co-workers, Barra was warm and unassuming, with kind eyes and an easy smile, but she also had a no-nonsense approach. Once while running a factory in Detroit, she told a union official that she wanted to get rid of the outside Christmas lights to cut costs. The union official was peeved. They eventually agreed to split the electricity costs.[28] Later, as head of human resources, Barra condensed the company's 10-page dress code to two words: "Dress appropriately."[29]

Barra's impatience with GM's plodding culture got her noticed by a new CEO who arrived in 2010: Dan Akerson. A gruff Navy-man-turned-telecommunications-executive, Akerson had been on the federal task force assembled to save the auto industry from the Great Recession. He knew nothing about the auto industry. But he quickly developed a distaste for what he viewed as an insularity and arrogance in the car business. "We know best. We're car guys!" he once said mockingly during a speech, thumping his chest.[30]

Akerson saw Barra as an antidote to that perceived machismo. He soon made her head of product development, the post that had been occupied by the legendary Lutz just 18 months earlier. A few years later, in 2013, Akerson advocated for GM's board to pick Barra to replace him. That vaulted the then-51-year-old from a relative unknown to one of the world's highest-profile corporate figures overnight. "The stodgy corporate culture at GM has changed forever," declared the New York Times.[31]

Early on, Barra sought to stamp out behaviors long blamed for a lack of accountability inside GM, and a reluctance to hash out conflict. She hired an outside facilitator to lead her executive team through weekend retreats that became a sort of group therapy sessions, where participants were encouraged to dredge up their own painful memories and confront one another about past conflicts.[32]

Barra also set about a strategy once unthinkable to generations of GM's think-big executives: she began shrinking the company. GM had been losing billions of dollars in Germany, India, Thailand and other overseas markets. How was that helping to underwrite the astronomical tab for developing electric and self-driving cars over the next decade? She began methodically pulling out of unprofitable markets, culminating in the unloading of GM's nearly century old European business, Opel, in 2017. That left GM as the only major automaker with virtually no presence in Europe.

"We've had to make some tough decisions and move away from trying to be everything to everyone, everywhere," she would say later.[33] Two months after the Opel sale, GM was ready to give a sneak peek of its plan for a flexible EV platform—a common set of battery cells, electric motors, and other mechanical guts—that could serve as the foundation for everything from a sleek sports car to a beefy pickup. GM summoned the media to its storied design dome in Warren. The silver, mid-century modern dome, designed by Finnish American industrial architect Eero Saarinen (who also designed the Dulles Airport terminal in Washington, DC, and the Gateway Arch in St. Louis), is where generations of GM executives and directors reviewed new vehicle designs, from muscular Corvettes to big, long Cadillacs with gaudy tail fins.

Journalists were greeted by Mark Reuss, then GM's product chief, a sharply-dressed executive with a deep, baritone laugh and a mercurial mood, quick to darken at the faintest whiff of GM criticism. Reuss is GM royalty and bleeds GM blue. His dad was the company's president in the early 1990s. With the departure of Bob Lutz—who finally retired in 2010—Reuss inherited the unofficial role as GM's resident car guy. He is pure Detroit, a gasoline-in-the-veins guy, certified to test-drive cars on Germany's famed and harrowing thirteen-mile Nürburgring.

That's why his message that day was beyond jarring. "General Motors," Reuss declared to the gathering, "believes the future is all-electric." He stood amid more than a dozen future electric models, secretively shrouded in silk sheets.[34] He was there arguably to tout the industry's most ambitious EV plans.

Some executives on her team weren't buying into the aggressive EV push, and questioned how the company could make money on EVs, given the high cost of batteries. During one tense meeting in 2018 meeting, Barra angered some on her team by killing plans for a high-end, flagship Cadillac that was due to get a roaring V-8 engine. The automotive press raved two years earlier when GM revealed the concept car, called the Cadillac Escala, but Barra thought going with a gas engine at that point would send the wrong message.

Work continued behind the scenes until March 2020, when Barra and her team were ready to show the world that they were betting the company on electrics. GM invited hundreds of dealers, investors, and Wall Street analysts to gather again under the soothing soft lights of the design dome. It was just days before the spreading coronavirus would shut down much of the world.

Barra strolled the floor as visitors ogled a dozen future all-electric models, some several years from seeing the inside of showrooms. In the car business, where future products are cloaked in secrecy, it was unheard of to show a slew of prototypes so far in advance. The vehicles spanned from mid-size Buicks to massive SUVs and even a sleek Cadillac that one executive said would be hand-built and priced north of $200,000. GM said the platform on which all those models would be built was to be called Ultium, a system of pouch-shaped battery cells that could be stacked vertically or horizontally below the floor for flexibility to accommodate all those size and body styles.[35]

Attendees left impressed. But GM's stubbornly sluggish stock price barely budged. Within weeks, it would plunge as the pandemic forced GM and the rest of the industry to shut down their plants. Wall Street analysts assumed the cash crisis would force GM to back-burner its investment in EVs—those sorts of future bets weren't going to help keep the lights on. Instead, Barra and her team doubled down.

In June, the CEO sat with Reuss and other top executives, again inside GM's design dome. But this meeting wasn't a design review. The executives sat around a large table, wearing surgical masks, to decide which future vehicles were on the chopping block. Details of the plans for each car—from minor facelifts to major new models—were spread across large digital wall charts, detailing potential launch dates and sales targets. The executives canceled some programs and scrapped others altogether. But by the end of the meeting, all the electric-vehicle projects on the board emerged untouched. A nearly $3 billion renovation of the Hamtramck factory to build them would also move forward.

"The situation allowed us to look at things with a very clear eye," Barra would say later.[36]

By fall 2020, it became clear that GM and its automaker peers not only would survive the industry's unprecedented shutdown but would bounce back faster than just about anybody expected. It turned out plenty of Americans wanted to buy new wheels despite pandemic uncertainty. Flush with savings from pausing other spending during the quarantine, car buyers were splurging, opting for pricier models and fancier trim levels. In the span of a few quarters, GM would go from gut-wrenching losses to record profits.

And another strange thing happened in the pandemic: EVs took off. Or more precisely, EV stocks took off. Shares of Tesla in 2020 went into ludicrous mode. By the end of the year, Tesla was worth more than $700 billion. Startup EV companies seeking to be the Tesla of trucks, of vans, of commercial trucking were turning investors' heads and raising billions.

Amid that frenzy, investors finally were willing to give GM a look. And Barra's team was ready with a steady diet of pronouncements about GM's electric ambitions. The stock rallied 5 percent when GM announced that it aimed to phase out gas-powered cars by 2035, the most aggressive timeline of any global car company. Investors applauded again when GM boosted its planned spending on EV development, to $35 billion. An announcement about a new electric-van service juiced shares 6 percent.

GM's stock had been stuck in neutral for the entirety of Barra's tenure. Even as she turned GM into a profit engine—surpassing historical returns and besting most of its global car-company rivals—investors yawned. Quarter after quarter, GM blew past Wall Street profit forecasts. And almost every time, investor response was muted at best. Finally, the perception of an aggressive EV strategy was stoking a cathartic rally. From an all-time low of $18 at the onset of the pandemic, GM's shares surged to $63 by that day Biden paid a visit to Factory Zero and felt the power of the gleaming white electric Hummer.

It would prove to be a fleeting moment of unrivaled optimism. Barra soon would be on the defensive about her company's slow-motion EV launches. GM's share price would begin to shrink, as did its perceived lead on EVs. It took one vehicle, from one company, to dent the perception of GM as EV leader. And that vehicle came from the most gnawing of rivals.

2

In China's Rearview

I n December 2011, several years after GM had crushed the last of its EV1s, a former GM employee named Bob Galyen found himself in the coastal megacity of Shenzhen, China.

Galyen was president of the battery division at Magna International, an automotive supplier. Tall and lanky, his red hair giving way to some gray as he then entered his late fifties, the Indiana native had led battery development for the EV1 but was cast off from GM when its massive parts division, Delphi, was made into its own company. Delphi struggled to stand on its own; the stock sank, and Galyen lost two-thirds of his retirement nest egg. He eventually landed at Magna, which brought him to China to give a presentation on EV batteries.

Walking off stage, he was approached by a man who introduced himself as Robin Zeng. An entrepreneur and chemist in his early forties, Zeng was then CEO of Hong Kong–based Amperex Technology Limited, or ATL, one of the world's largest suppliers of small lithium-ion batteries used in laptops and other consumer electronics. Also known as Zeng Yuqun, he was a high-profile Chinese businessman at the time, but Galyen had never heard of him. Galyen's expertise was in lithium-ion batteries for EVs, not consumer electronics.

Zeng said some of his engineering managers wanted to meet the American battery exec, who they considered a sought-after expert on the use of lithium-ion batteries in EVs. Zeng invited Galyen to have dinner near the conference.

Galyen thought it would be an intimate dinner with a handful engineers. Instead he walked into a room with six tables, each situated with ten neatly placed chairs. The centerpiece of each table was a framed copy of an issue of *Batteries International* magazine with Galyen's picture on the cover.

Galyen was stunned. He turned to Zeng, who laughed and explained that everyone in the room had read the article and was eager to hear more from this battery guru from the US Midwest. "I want you to inspire my young people to be better at what they do," Zeng told him.[1]

Galyen smiled but still felt blindsided, not quite sure what wisdom Zeng was looking for him to impart. "OK. Let's have dinner," he said. The self-proclaimed simple Indiana farm boy found himself feasting on a twelve-course meal with sixty Chinese chemists and engineers, answering their questions and spinning tales of working on GM's storied EV1. They dined on fresh-caught seafood, and later washed it down with shots of Maotai, the clear, potent sorghum-based liquor that is revered in China, having been served on official state visits, from presidents Richard Nixon to Barack Obama.

The next day, Zeng invited Galyen to visit his battery factory, more than an hour's drive away, in Guangzhou Province. At a conference room table with directors and VPs huddled around, Galyen disassembled some of ATL's battery cells, scribbling notes. The cylinder-shaped batteries were much smaller than the EV cells he normally worked on. Galyen tinkered, offering tips on how they might improve their efficiency and design. More engineers filtered in. Before long, two dozen or more were thronging the table.

Galyen emerged from the conference room about the same time Zeng had returned from attending to an urgent matter elsewhere in the building. Zeng seemed surprised at Galyen's impromptu battery workshop: "Why did you do that, Bob?" he asked. The US executive shrugged. "Well, you're not in a competing industry," he said. "I'm building electric-vehicle batteries. You guys are building consumer-electronic batteries. I don't care if I fix your problems because you're not competing with me."

Zeng grinned.

Afterward, Galyen lunched with some ATL investors, along with its board chairman, Mike Chang, a Notre Dame PhD in electrical engineering who spoke fluent English. Battery technology dominated the discussion. Galyen hit it off with Chang and everyone else.

At the end of the day, Zeng finally told Galyen that in fact ATL was considering a new division to make batteries for electric cars. He wanted Galyen to run it. "Would you ever consider coming to work in China?" Zeng asked.

For the second time, Zeng caught Galyen off guard. "Well, maybe?" he stumbled. The two men agreed to meet next time Zeng was in the United States, months later, when Zeng would be in Cupertino, California, for a meeting with Apple. Zeng was there to land what would become ATL's gargantuan main customer, supplying cells for Apple's iPhone, which at the time was surging in popularity across the globe.

For Galyen, the offer loomed as an unexpected left turn in what had been a long, distinguished career in batteries. Galyen knew it wasn't a typical move to make in the twilight of a career. He had visited China often on work trips but had never envisioned moving there—especially not for a company he'd never heard of. But if anyone could pull off such a career move at a stage when most are mapping out their retirement plans, it was Galyen, a man whom one colleague described as a "whirlwind of energy."

"What initially went through my head," Galyen recalled, "was that my wife would probably be angry that I was thinking about taking a job in China."

Galyen grew up in Knightstown, Indiana, known as the setting for the 1980s high school basketball movie *Hoosiers*. Galyen didn't play hoops, though, despite his six-foot-three-inch frame. He was a track star—his school high-jump record, of six feet six-and-a-half inches, stood for decades,. Galyen's father worked as a farmer on state-owned land, tending dairy cows and planting crops. His mother stayed at home until his father had a heart attack and couldn't go back to his job. She took a job at a local casket-making company, which at the time was busy making coffins for soldiers killed during the Vietnam War.

One of his earliest memories was his dad explaining that some birds were threatened with extinction because chemicals in the environment were making their eggshells too thin, and they were cracking when the birds sat in their nests. "He said, 'We've got to stop putting these chemicals in our environment because it's destroying our ecosystem,'" Galyen recalled. "My father wasn't wrong."

Galyen is a different breed of environmentalist, a term he doesn't identify with. His view of the mainstream environmental movement is dim—he calls activists "tree huggers." "OK, what are you guys doing to save the planet?" he says in a measured but slightly exasperated way. "You're talking about it? I don't talk about it. I just go do it."

That's why Zeng's offer was so intriguing. Galyen had spent decades working on battery development in the United States, mostly at GM. He had several battery-related patents to his name. And yet, at the time of Zeng's offer in 2011, the United States was basically no closer to electrifying its transportation network than it was before his career began. He watched GM make breakthrough after breakthrough on electrification, only to pull back, to change course, to literally crush his dreams. Now he was being given the chance to make a big difference—fast.

"Here's a chance to use the skill sets that I had been developing for three decades and turn it into something that's going to help bring value to mankind," he said. "It wasn't a mission to help China. It was a mission to help save the world." He took the job.

By July 2012, Galyen was sitting in his office in Ningde, a mountainous fishing town, across the strait from Taiwan, known for its tea plantations and aquaculture farms of oysters, prawns, and freshwater eel. ATL created a sibling company called Contemporary Amperex Technology Limited, or CATL. His recruiter Robin Zeng was employee number one. Galyen was number two.

In less than a decade, CATL would be the world's dominant player in EV batteries. It would rocket Zeng's net worth to more than $30 billion, making him one of the world's wealthiest people and among China's highest-profile businesspeople.[2] For Galyen, an unexpected encounter at a conference would vault him to the pinnacle of his career, a lucrative stint that would finally give him the satisfaction of scaling a battery enterprise while making a difference on the climate crisis. It was somewhat controversial, though, as Galyen was seen by some as helping to catapult China into a commanding lead in the global EV race.

. . .

"Lithium batteries are the new oil," Elon Musk tweeted in July 2022. The billionaire entrepreneur is prone to provocative overstatements, but this particular sentiment is shared by many economists, foreign-policy experts, and politicians. If it's true, though, then China is the new Texas, Saudi Arabia, or Russia—or maybe the three of them combined.

In the span of less than two decades, China has morphed from the go-to manufacturer for the world's cheap consumer goods to the gatekeeper of the entire electric-car value chain. China's dominance in electric vehicles and batteries is hard to overstate and is driving the inevitable, but messy transition away from internal combustion engines. By 2022, the country accounted for about 60 percent of all electric-car sales. It owned more than three-quarters of all battery-making capacity. A majority of the raw ingredients needed for lithium-ion battery cells is processed in China; more than half of the world's lithium and nickel is refined there, nearly all of the manganese.[3]

A relatively small amount of that stuff is actually mined in China. Instead, it's extracted in places like Chile, Indonesia, Australia, and the Democratic Republic of the Congo (DRC), where Chinese companies own vast swaths of those operations. Cobalt mines in the DRC, for example, account for 70 percent of the world's supply of the silvery-blue metal used in the most common EV battery chemistry—and Chinese companies own 80 percent of the cobalt mines in Congo, where Chinese lettering adorns casinos, restaurants, and hotels in the mining belt in the southern part of the country.[4]

This dominant position China has seized so early in the global scramble for EVs and batteries is daunting if you're an auto executive in Detroit or Tokyo, or a politician in DC or Brussels. A well-timed gamble by Beijing that China could leapfrog the world's richest, most industrialized countries to win the global race for the transportation revolution of the twenty-first century has paid serious early dividends. A confluence of factors put China in the pole position. Some of its advantage is the result of a central-planning regime that only Communist China can pull off through sheer force and gobs of government money. Regardless, China's abrupt emergence as an EV powerhouse is starting to have far-reaching geopolitical implications, as the world's largest country exerts outsized influence over the key building blocks for how many human beings worldwide will travel in the decades to come.

Its leadership on batteries is especially surprising given that rechargeable lithium-ion batteries were developed in the 1970s and '80s by three chemists from America, Britain, and Japan: John B. Goodenough, a longtime University of Texas chemistry professor who died in 2023; British-born M. Stanley Whittingham, a faculty member at the State University of New York at Binghamton; and Akira Yoshino, a chemist at a major Japanese chemical company. The three were awarded the Nobel Prize in 2019 for their work refining the technology that has enabled the proliferation of wireless devices like iPhones and laptops, and now, vehicles. The rechargeable, lithium-ion battery "makes a fossil fuel-free world possible, as it is used for everything from powering electric cars to storing energy from renewable sources," the Nobel committee wrote.[5]

Batteries had been around for more than a century before the lithium-ion variety arrived. The heavy lead-acid batteries long used in automobiles have relatively low energy densities, enough to turn over the engine and power the lights and radio, but not nearly adequate to propel the car. For decades, consumer electronics like handheld video games, camcorders, and cordless phones were powered by nickel-cadmium batteries, which since the nineties have given way to lithium ion, to the point that they are now relegated to the scrap heap of battery history.

Lithium-ion batteries are better than predecessors in almost every way. When a lithium-ion battery is discharging, or powering a device, lithium ions pass from the negative electrode, the anode, to the positive one, the cathode. That process also sends electrons from the anode to the cathode via a circuitous route, an outer circuit. When the battery is charging, the opposite happens: lithium ions are released by the cathode and stick in the anode. That's the part the user notices—the little lighting icon when your smartphone is powering up.

Since it all starts with an electron leaving its home in the anode, the ideal material for an anode is one that easily lets go of its electrons. Lithium, the Nobel committee would state in awarding the prize, has an "enormous drive" to release its outer electron.[6] And, since it's the lightest of all metals, it has the benefit of packing the most energy for its weight.

Whittingham was a Stanford chemist who went to Exxon in the early 1970s, around the same time the oil embargo was prodding Exxon and other oil giants to explore alternative energy sources. He understood lithium's potential and was tinkering. Whittingham paired a lithium anode with a titanium disulfide cathode, a yellowish metal that had never been used before in batteries. Its molecular structure created tiny pockets that allowed the lithium ions to burrow in. The result was a battery that stored roughly ten times the amount of energy as lead-acid batteries and five times more than nickel-cadmium.[7] The only problem was, repeated charging made those early batteries prone to exploding. Whittingham's team worked to further develop the technology, but a drop in gasoline prices later in the 1970s led Exxon to mothball development.

Goodenough took the baton in 1980, while working as a professor at the University of Oxford in the UK. The chemist had a hunch Whittingham's titanium-based cathode could be upgraded. He and others tested a cobalt-oxide cathode. Boom. It generated nearly twice the voltage. "This," the Nobel committee concluded in 2019, "was a decisive step towards the wireless revolution."

Yoshino provided the third leg of the lithium-ion battery's discovery by finding a way to remove pure lithium from the battery itself, leaving only lithium ions passing back and forth through the electrolyte. This made the batteries far less prone to blowing up, and thus commercially viable.[8] Starting in the 1990s, Sony and other Japanese electronics companies began switching to the lightweight, compact lithium-ion batteries for use in millions of portable music players, camcorders, and handheld gaming devices. Smartphones would follow.

Early on, Korean and Japanese battery makers, including Panasonic, ruled lithium-ion battery production, accounting for nearly 90 percent of global output. By the mid-1990s, though, China had begun investing in lithium-ion factories and development centers. Private companies sprang up, including Robin Zeng's company, ATL.

Another major battery maker was born quietly around the same time: BYD, based in Shenzhen, like ATL. Around 2000, BYD landed a contract with Motorola as the cell phone giant's first Chinese provider of lithium-ion batteries.[9] China's rise as the world's battery powerhouse was on.

· · ·

China began opening the country to foreign investment beginning in the late 1970s, after several decades of virtual economic isolation. At that point, roughly two-thirds of China's population lived in poverty.[10] Free-market reforms attracted investment from the United States and European countries. The influx of overseas capital, combined with massive amounts of domestic spending by the Chinese government, helped the economy grow an average of roughly 10 percent annually for four decades, through 2017. The World Bank called this "the fastest sustained expansion by a major economy in history," helping to lift hundreds of millions out of poverty.[11]

The automotive sector was a centerpiece of Beijing's industrialization effort. China's car-making ambitions date back to the 1920s, when Sun Yat-sen, who was the first president of the Republic of China after the overthrow of the Qing Dynasty, wrote a letter to Henry Ford. He urged the auto baron to consider an exploratory visit to South China.

"I know and have read of your remarkable work in America," reads the typed letter, dated June 12, 1924. "I think that you can do similar work in China on a much vaster and more significant scale." Ford's office responded nearly five months later with a two-sentence letter that stated: "Mr. Ford has made no plans for visiting China in the very near future."[12]

It would be seventy more years before China's auto sector stirred to life. Private-vehicle ownership in China was mainly reserved for government officials and dignitaries through much of the twentieth century—Sun Yat-sen was fond of GM's Buick brand, which oddly became a favorite among China's elite. Once Communist rule began in 1949, foreign brands were cut off. Small-scale auto manufacturing began in the 1950s with some investment from the Soviet Union. By the early 1990s, there were only about 5.5 million civilian-owned cars on the road in China for a population of around 1.1 billion people—roughly one car for every 200 people.[13] The United States by comparison had about 134 million cars on US roadways—nearly one car for every two people.[14]

For developed countries, auto industries during the twentieth century shaped national identities. The mention of German engineering conjures up Volkswagen and the country's luxury juggernauts, Mercedes-Benz and BMW. The United States has the Detroit Three of GM, Ford, and Chrysler (today it's more like the Detroit Two-and-One-Third, with Chrysler having been absorbed into Stellantis, a French-Italian-American melting pot that owns the Jeep and Ram brands). Think of Japanese industry, and automotive powerhouses like Toyota and Honda come to mind.

By the 1990s, Chinese government leaders wanted to flex the country's fast-growing manufacturing might in the car business. But unlike how the United States and European nations grew automotive industries by planting flags in new markets, China would entice outside investment by requiring foreign car companies to partner directly with Chinese automakers, most of which were fully or partially state-owned.

For the foreign automakers, these joint-venture (JV) arrangements offered a foothold in what would become the world's biggest car market. Foreign companies got capital to build factories, and the Chinese JV partners helped them navigate the Communist regulatory system and market intricacies.

For China, the arrangement would create a surge of new jobs and businesses, as a huge injection of capital spawned factories and built out supply chains. It also helped seed automotive and manufacturing expertise in China, as the fledgling Chinese automakers learned from their foreign partners.

This JV movement set off a gold rush among car companies. In the mid-nineties, GM and Ford were locked in a frenzied competition to court Shanghai Automotive Industry Corp., or SAIC, one of the largest and most sophisticated of China's automakers. For more than a year, Ford stationed about a dozen employees at a hotel in Beijing to manage the proposal. The team converted hotel rooms into cramped, makeshift offices, with bathtubs serving as filing cabinets.[15] The contract, though, ultimately went to GM.

"It has redefined what a great growth opportunity is," GM's then-CEO Rick Wagoner said of China in a 2004 Bloomberg interview.[16]

Volkswagen also secured a JV with SAIC. With the help of the Shanghai-based company, those two giants—VW and GM—quickly became China's biggest car sellers. By the early 2000s, China was the main strategic focus of just about every major global automaker. Near the end of that decade, VW's vehicle sales in China surpassed its sales in its home market in Germany.[17] GM sales in China would eclipse its US sales the next year.[18]

Hundreds of domestic Chinese manufacturers also emerged, most targeting the vast, lower end of China's burgeoning market, selling new vehicles under $10,000. These cheap cars gave urban buyers an alternative to the ubiquitous scooters that buzzed around China's megacities. It also offered greater mobility to peasants in the countryside who needed basic transportation. Other Chinese brands, like Geely and Chery, were larger and more sophisticated, aiming to compete in the mass market against GM, VW, Toyota, and other multinationals.

Big or small, many of the Chinese automakers in the early twenty-first century had one thing in common: they produced cars that very much looked like their counterparts from established players. From the floor of the Shanghai auto show in 2007, *MotorTrend* auto writer Mike Floyd filed this dispatch: "Like the fake Rolexes and Fendi bags hawked by the city's swarming street vendors, there were plenty of cars on display that looked all too familiar on the surface."

Still, Floyd's write-up noted an undeniable vibe in the show hall. "Despite all the imitation, you could feel the energy in the convention center, and sense that the Chinese auto industry is on the verge of breaking out and becoming a true worldwide player."[19]

At the time, though, the local Chinese car brands simply couldn't compete on internal combustion engines. The multinational automakers had a century head start on power-train engineering. Chinese consumers viewed brands like VW, Toyota, and Buick as more reliable, more refined big brothers to Chinese brands. It was a shortcoming not lost on Communist Party leaders. There was a growing sense among government officials that Chinese carmakers might never close that gap in quality and sophistication.

. . .

In 2007, China's central government picked Wan Gang for the prestigious post of minister of science and technology. It was a curious choice. Gang, a Shanghai native and mechanical-engineering PhD then in his mid-fifties, had been an Audi engineer. He was the first non-Communist Party member to be put into the job, and his appointment underscored China's EV ambitions.

For years, Gang had been involved in electric-car research, and most of the time he'd been whispering in the ear of every Chinese leader who would listen about the potential for EVs. Bypassing internal combustion engines for electrics could not only help China catch its bigger rivals in the auto industry, but surpass them, Gang would tell them.[20] Seven years earlier, in 2000, he had submitted to China's State Council—roughly equivalent to the US president's Cabinet—a proposal titled, "Regarding Development of Automobile New Clean Energy as the Starting Line for Leap-Forward of China's Automobile Industry."[21]

Gang would remain minister for more than a decade and is credited by many to be the architect of China's EV strategy. He would eventually cross paths with another notable EV advocate: Elon Musk. In 2008, Gang visited Tesla's headquarters in San Francisco. Some of Gang's friends had encouraged him to visit the American entrepreneur whose EV startup had been getting some media buzz. Inside a giant tent, Gang test-drove a green Roadster EV, Tesla's first vehicle, which went on sale that year.[22] The men would connect with each other over the years at various industry gatherings around the globe. Gang seemed to view Musk as a frenemy: he figured Tesla would eventually score big in China and challenge local companies—and it did. But Gang pointed to Musk's moves, including manufacturing and sales strategies, as examples for fledgling Chinese EV makers to follow.

In 2009, China's government made its big bet on EVs official during an address in Beijing at the ornate Diaoyutai State Guesthouse, a sprawling, resort-style hotel akin to the US's Camp David. Party leaders outlined ambitions for China to become the world leader in electric cars. In just two short years, it was declared, so-called new-energy vehicles, which included hybrid gas-electric cars, would account for a half-million of China's new cars produced.[23]

There were three main reasons for China's decision to bet on electric cars. One was to wean itself from dependence on foreign oil. The second was, by cornering the market for electric cars and batteries, there was the potential to create vast new supply chains and hundreds of thousands of jobs.

The third was less opportunistic and born more from desperation. Air pollution in China's megacities had become a dire problem and a major embarrassment for the country's leaders. On a 2007 list of the world's twenty most polluted cities, sixteen were in China.[24] In Beijing, dust and soot mixed with sand blown in from its nearby desert to regularly shroud the city in a hazy smog. When Gang advocated for EVs, he often paired the economic argument with his belief that the country needed to reverse its slowly unfolding environmental calamity.

"You can chew on the air in most cities," media executive and longtime Beijing resident James McGregor told the Voice of America radio network in 2009. "The rivers are Technicolor with effluents. This place has gone through a huge economic boom, and they've just been ignoring it."[25]

The government tried a last-ditch cleanup ahead of the 2008 Summer Olympics in Beijing, aware that the environmental mess would soon be on display for all the world. Athletes were voicing their concerns and plans to wear masks, with some threatening to sit out the games altogether.

China spent an eye-popping $10 billion on remediation efforts in the years leading up to the opening ceremony.[26] It shut down the worst factories and took old, coal-burning furnaces offline. It planted trees. It restricted other cars from entering the city on odd or even dates during certain periods, depending on their registration numbers. On a Saturday in September 2007, the government held a voluntary "no-car day," urging motorists in more than a hundred cities across the country to stay off the roads.[27] Various programs that offered cash for old beater cars, both before and after the Olympics, removed millions of soot-spewing gas or diesel engine vehicles from the roads.

China deployed an elaborate regulatory system of sticks and carrots to cajole carmakers into producing EVs and entice consumers into buying them. It offered billions in subsidies for EV output. The government in some cases all but guaranteed the companies that there would be buyers. Beijing set aside $1.3 billion to replace some 70,000 gas-powered taxis, for example.[28] Consumers, meanwhile, could combine national and local government incentives to knock off as much as half the price of an electric car. The *Wall Street Journal* in 2017 quoted a dealership manager for Beijing Auto as saying his bestseller was a petite, snub-nosed electric SUV priced at $22,000—but routinely went for $11,000 after government subsidies.[29]

Restrictions against ownership of gas-powered cars in its largest cities grew more stringent. At one point, an average of 11 million people were applying every month for one of 14,000 licenses for gas-powered cars in Beijing—785 applicants per license. They would go to the highest bidder. EV buyers could get tags with no red tape—at no cost.[30]

In a way, the Chinese were quietly making the same wager that Elon Musk was: that their competition wouldn't, or couldn't, seize the emerging market. The world's traditional automakers, without being pressured, could dabble in EVs, but they weren't about to wholeheartedly embrace electrics. Their bottom lines were beholden to gas and diesel engines—for the Germans, big, sporty luxury sedans; for Detroit, honkin' pickup trucks; for Japan, sensible, reliable fuel sippers. Both Musk and Chinese leaders were placing separate, massive bets that none of their competition had the incentive or the guts to detach themselves fully and completely from fossil fuels.

. . .

In November 2011—just a few weeks before Bob Galyen would toss back shots of pungent liquor with Chinese battery chemists at ATL—Elon Musk was on TV with a Bloomberg journalist. Musk's team was preparing the US launch of its second car, the Model S. Musk was asked his opinion of BYD, the Chinese upstart that had been racking up electric-bus sales in China and the United States, and also starting to sell quite a few EV passenger cars in China. BYD was unknown in the US, but Musk knew about the company.

The Tesla CEO grinned, then broke into laughter. It wasn't just a dismissive chuckle. It was a full-blown, Dr. Evil-worthy belly laugh, head tilted toward the ceiling.

The journalist interjected: "BYD is trying to compete. Why do you laugh?"

Musk composed himself and responded, still grinning: "Have you seen their car?" Pressed further, Musk explained that he didn't see BYD as a rival at all. "I don't think they have a great product. I don't think it's particularly attractive. The technology is not very strong."[31]

Musk's scorn aside, others who knew this Chinese company—which by then had applied the fanciful, maybe goofy "Build Your Dreams" acronym to its name—were taking it more seriously, including famed investor Warren Buffett, who had taken about a 10 percent stake in BYD.

BYD was started in the mid-1990s as a battery company. It was founded by Wang Chuanfu, a chemist who was the son of poor farmers in rural China. He was raised by his older siblings after his parents died.[32] Chuanfu went on to get a master's degree in chemistry in Beijing. He spent the early part of his career as a government researcher before striking out on his own.

Chuanfu's secret genius was an unusual ability to reverse-engineer other people's products—not simply to rip them off but make them better while also making them cheaper. In his early years, he closely studied the Japanese battery juggernauts, like Sanyo and Panasonic, to this end.

Much of BYD's early commercial success is credited to one of Chuanfu's earliest recruits, Stella Li, a scrappy twenty-something when she joined the company as marketing chief after graduating from Fudan University, one of China's most prestigious schools. Within three years, Li had opened BYD offices in Hong Kong and Europe, and began targeting some of the world's largest electronic-device makers for potential business.[33]

She made many trips to Motorola's headquarters outside Chicago. On her first visit, she bribed security guards with doughnuts to get access to key purchasing engineers and prove to them that BYD's lithium-ion battery cells were every bit as good as the ones they had been using—and far cheaper.[34] Within a few years, BYD landed a multibillion-dollar contract with Motorola and a similar deal with Nokia.[35]

BYD's early success spurred Chuanfu to expand into the automotive business—but not just the car *battery* business. He wanted to build cars. When the company announced in 2003 that it was acquiring a struggling maker of passenger cars and buses, BYD's stock tanked. Investors didn't like the sudden focus shift for a company that already had its hands full with massive orders from industry giants. Chuanfu didn't blink, applying the same aggressive expansion strategy to his auto business as he did the battery division.

Within six years, BYD had the top-selling sedan in China and had started making plug-in hybrid vehicles, a sophisticated technology that even established automakers had barely dabbled in. By 2008, BYD was approaching a half-million in annual car sales and had begun exporting its cheap vehicles to South America and the Middle East. Buffett's firm that year would invest about $230 million in BYD.

Chuanfu was so optimistic about his car company's prospects that he took BYD to the Detroit auto show, one of the big-boy events on the car-show circuit. BYD was known then as one of the largest makers of rechargeable batteries in the world, but its car division was more likely to draw snickers of laughter than praise from the thousands of auto journalists who flocked to the event.

Chuanfu and other BYD execs presented a new hybrid system during a press conference there; Later, the chairman hopped in the car with a reporter from the car-enthusiast site Jalopnik to show it off. To the surprise of just about everyone, Chuanfu took off, driving the thing with the reporter in the passenger seat, weaving through carpeted stretches of show floor, at one point through an active press conference for the American Le Mans Series. The episode, as Jalopnik reporter and passenger Matt Hardigree described it, was "illegal and surreal," and amounted to "breaking all the rules of proper auto show behavior."

Still, he added: "That electric motor? Quiet as a mouse."[36]

In the years following that rogue test drive, BYD emerged as a powerhouse manufacturer of electric buses—selling many in its hometown of Shenzhen. During the 2010s, the city put into service some 16,000 electric buses and 22,000 electric taxis, many from BYD.[37] Before that binge, the city had zero electric buses. For comparison, the City of Los Angeles's entire bus fleet—electric or otherwise—in 2023 totaled about 2,300.

Buffett's top lieutenant, Charlie Munger, would say years later of Chuanfu: "He can solve technical problems like Edison and get things done like [former GE CEO Jack] Welch. I've never seen anyone like him."[38]

In 2022, as companies like GM were targeting a 2035 date to completely switch over to electric cars, BYD said it was making the move immediately. In the fourth quarter of 2023, BYD eclipsed Tesla as the world's biggest EV maker. Musk wasn't laughing anymore. "Their cars," he said in a May 2023 tweet, "are highly competitive these days."

BYD emerged as the most successful of the homegrown Chinese automakers, but hundreds of other EV startups were popping up there, all betting that the barriers to entry had fallen far enough that they had a shot. The vast majority would never turn a profit and were destined to fail. But a handful would rise to challenge the VWs and GMs in the world's largest car market, and beyond.

. . .

For an auto executive, media day at a global auto show is an important event, and a reunion of sorts. You bump into former colleagues, old rivals, and suppliers and entrepreneurs, eager or desperate to court your business. It's an espresso-fueled spectacle of professional networking, fashionable suits, and eye-catching sheet metal. As far as professional conferences go, there are worse places to spend a day.

At the Shanghai motor show in 2015, the buzz was maybe even more pulsing than the other high-profile stops that marked the carefully laid-out auto-show circuit: Detroit, Geneva, Paris, Frankfurt, New York, Los Angeles, Tokyo. The big guns were in Shanghai in full force—Mercedes and GM both showed driverless concept cars that looked like sci-fi movie props. Tesla showed a version of its Model S sedan with its head-snapping speed.

There were also some forty Chinese automakers, showing everything from knockoff Jeeps to cheap electrics.

That year, the Shanghai Auto Show was held in a massive new convention center, sprawling with nearly sixteen million square feet of space, one of the largest buildings in the world. The concrete had barely dried in the place, but the government had decided that the annual show deserved an upgrade and rushed to finish it for the show. The building was laid out in a four-leaf-clover design, meant to make the walking efficient. But a visitor to the car show still would typically walk miles on carpeting and concrete walkways in between the giant, glitzy displays.

Jack Cheng stood in a trim, dark suit near one of those glitzy displays—nearby a $600,000 red Ferrari 488 GTB shining under gleaming lights. Cheng was running vast swaths of the Asian business for Fiat Chrysler Automobiles, or FCA, the Italian American carmaker that then owned the Ferrari supercar brand (and would later merge with France's Peugeot S.A. to create Stellantis). His boyish good looks, quick smile, and penchant for jamming Led Zeppelin songs on his guitar at company functions made him seem younger than his age, then in his late fifties. He was here showing off sexy supercars in what he figured might be the final chapter in his decades-long career in the car business.

While holding court at Ferrari's stand, Cheng was approached by an entrepreneur named William Li, about two decades Cheng's junior. Li grew up raising cattle with his grandparents in a mountainous region west of Shanghai and later attended the prestigious Peking University, where he got into computer programming. He made his money building and then selling a company that made software for car companies and dealers. Now, he was interested in making the cars himself. Electric ones. And he wanted Cheng to help with the car-making part.

"We're all tech guys. We don't know how to build cars. You do,'" Cheng later recalled Li saying to him. "Why don't we start this new company?"[39]

Cheng was wary. He'd seen plenty of Tesla wannabes. Most were little more than vaporware. "I was so skeptical about these son of a guns. They're nothing but money," Cheng said. "I told him, 'Do you have any idea how much money you'll have to spend in the car industry?'"

Cheng decided to call one of his sons in New York who had spent time at a big consulting firm. His son huddled with a few of his former colleagues for an impromptu research session on the Chinese auto market and found growing investor appetite for exposure to China's budding EV industry. They also saw that Li's nascent company—then called NextEV—had backing from Chinese tech heavyweight Tencent Holdings, the world's largest video-game maker and then owner of the WeChat messaging app. Tencent at the time was on its way to a stock market valuation of nearly $1 trillion, one of the world's largest.

Two days later, his son called Cheng back. "Do it," he told him. "Quickly."

. . .

Cheng was raised in Taiwan, the son of a shopkeeper. When he went to college to study engineering, he figured it would enhance his chance of meeting girls if he learned to play the guitar. He taught himself by ear, listening to songs from Don Henley and Glen Frey of the Eagles on an old .45 record player. "I had to try like a hundred times before I could play 'Hotel California,'" Cheng said.

Unlike his parents, he wanted to see the world. He landed an internship at Ford China but was determined to get to the United States. He applied to colleges in Michigan, figuring it would hook him into the auto industry. He got accepted by Michigan State, but was rejected by his top choice, the University of Michigan. When Ford offered to send him to Michigan on assignment, he jumped at the chance and decided to skip school.

Cheng spent a few years in product-planning roles at Ford before taking overseas assignments in Australia and the UK. He eventually got sent back to China in the late 1990s to help set up Ford's growing network of factories there, and later ran its purchasing arm in the country.

One day in 2012, a headhunter emailed Cheng about interviewing with FCA. The Italian American automaker was strong in the United States with its Jeep and Ram truck brands, and in Europe with its Fiat cars. But it had yet to crack the Chinese market. Cheng showed up to a meeting at a posh Hyatt in Shanghai. FCA CEO Sergio Marchionne greeted him with some other executives at his side. Machionne was an industry sage, well-known for brandishing a half-dozen smartphones and a consistent uniform of dark sweaters, but almost never a suit. He was also known for his brutal honesty. He once summed up a flawed rollout of a new Maserati SUV to a magazine this way: "I think we sucked at the launch of the Levante."[40]

The men met in a smoking room on the eighty-sixth floor. Over ample espressos, they discussed the prospect of Cheng coming to help Marchionne get FCA's China business off the ground. The meeting must have gone well. As Cheng made his way to the elevator after it ended, an HR staffer chased him down. "When can you start?"

The extroverted native of Taiwan accepted the gig, and he quickly took to the Italians. They seemed to get that the car business was all about passion. They understood the power of brands. He liked how they were fashion-forward. He was into their trim-fitting suits. He remembers the suits he wore back when he worked at Ford in Michigan, oversized and unflattering on Cheng's small frame.

"Working in Detroit I was wearing very slack trousers, a big suit, like goddamn pajamas," Cheng said later. "The Italians? Everybody dresses like a celebrity. I started to realize there are differences in lifestyles."

For a couple of years, Cheng was happy at FCA, running its main parts division and its lending arm in the Asia-Pacific region. But he also could feel a change in the industry's direction. EVs and high-tech car features were taking off. Fiat Chrysler was a laggard on electrification. Marchionne used to joke—publicly and provocatively—that the company was losing $14,000 on every electric car it sold, so he urged his managers to not try too hard to sell them. His point was intentional: Marchionne wanted regulators from Milan to Sacramento and Beijing to know that the math on electric cars simply didn't make sense without massive government support.

At the time, FCA was trying to get a foothold for its juggernaut Jeep brand in China. The feedback from consumers was predictable, but with a twist, Cheng explained: yes, they thought the Jeep Grand Cherokee looked beautiful and rugged. "But, they would ask, 'Is this vehicle smart enough?'" Cheng recalls. "The key factor for purchasing a car in China had changed. It wasn't about driving performance. They wanted it to be smart. They wanted the tech. And EVs were this clean, easy platform to deliver the car as a smart device."

Chinese customers wanted extras such as voice commands and driver-fatigue alerts, even on low-priced models. They wanted fast upgrades in car tech, and they were getting them from Chinese brands like BYD and Geely, not so much from GM, Ford, VW, or FCA.

Cheng also started to see that this EV thing had a cool factor to it. It seemed these native Chinese brands, which for so long had been the little brother to the Western car companies, could take on the big brother with this tech.

Cheng broke the news to his boss, Mike Manley, a high-level FCA executive at the time. He was leaving to take a job with an unknown Chinese upstart, NextEV. To Cheng's surprise, Manley wasn't upset. Instead, he said, "Jack, you're going to make a lot of money."

William Li would rename the startup NIO and serve as the face of the company. Another founder, Lihong Qin, a Harvard grad who had experience in sales at a big Chinese carmaker, would handle the retail channel. Cheng would do what he had done for decades in the car business: set up supply chains and factories. But he was now thrust into the startup world, with none of the institutional backing that came with working at a multinational global powerhouse like Ford or FCA. He needed to build a supply chain from scratch.

A few months in, Cheng hosted a conference for parts suppliers to introduce its supply chain plans. Managers from about 250 suppliers arrived at a Formula One racetrack west of Shanghai. Many were old friends whom Cheng had worked with for years at FCA and Ford. A few were from some of the world's biggest auto suppliers. They were impressed when shown a clay model of a sleek SUV that looked a lot like a BMW. Cheng broke out the guitar, playing Guns N' Roses, along with a special song he had composed about NIO. The crowd ate it up.

But when the one-on-one conversations about contracts began, smiles faded. Agreeing to a long-term supply deal with an unproven startup car company of any sort was a magnitude of risk that parts suppliers were not used to. The chance of a supplier becoming an unsecured creditor in a bankruptcy case involving a startup Chinese EV maker was high indeed.

"They said, 'Jack, this is going to be very challenging for you to do this. There's just not enough volume'" to get the economies of scale typically needed to cover the capital costs, Cheng would say later. Even his old buddies from FCA weren't sold. "They said, 'Jack, we don't believe in you. This is going to cost you an arm and a leg,'" Cheng recalled later.

Cheng didn't try to persuade the supplier executives that they were wrong. He cut right to it: "How much money do you need?" making clear he was willing to pay a premium to help offset the risks of doing business with a risky startup.

Li wanted to grab the premium end of China's EV market. While Cheng worked on the back of the house, Li was out front, ensuring that the startup's cars would have sexy designs. Li had repeatedly shuttled to Europe, interviewing more than a hundred car designers before hiring Kris Tomasson from BMW in early 2015. Working from Munich, Tomasson's team grew to thirty people in three months as they quickly went about producing a foam prototype for a supercar.[41]

With Li fundraising and Cheng working on the supply chain, the company managed to pay off the development costs for the NIO EP9, a racetrack-only, two-seat supercar that would make a Lamborghini owner salivate. The EP9, with a price tag ranging from $1.5 million to $3 million, was a 1,300-horsepower monster that sprinted from zero to sixty miles per hour in 2.7 seconds.

That gave suppliers a degree of confidence in this young company with big ambitions. The NIO EP9 also put NextEV on the radar of the automotive press. But the company had bigger aspirations than a one-off exotic racer. It wanted to unlock the luxury market with a gorgeous, seven-passenger SUV that took inspiration from Audi, BMW, and other high-end powerhouses.

NextEV was renamed NIO, after the EP9's model name. The founders forged an ultrahip brand by opening dozens of so-called NIO Houses—showrooms, yes, but with the vibe and aesthetic of an indie coffeehouse inside a high-end spa. Many are designed by renowned local architects and found in posh retail centers across China and Europe. Some have reading rooms and even attached childcare setups.

NIO also sought to set itself apart from hundreds of Chinese EV start-ups by creating depots where customers could swing by to have their battery completely replaced in under five minutes. If you had more time, there would be NIO-branded charging stations. Soon, NIO earned the media's informal moniker of "China's Tesla killer."

Still, like most startups, it racked up losses as it scaled production—more than $3 billion in red ink from 2017 to 2019. Cash was dwindling, and its share price sank from $10 at its IPO in 2018 to less than $2, a period Li would later describe as "an extreme stress test." Wall Street analysts estimated that the company was within weeks of going under when it received a lifeline in the form of a roughly $1 billion financing package from Hefei, the capital city of Anhui, Li's home province. In an example of how Chinese EV players often enjoy government backing to a degree not afforded young western companies, the state would get a 24 percent stake in NIO, and the company agreed to establish a headquarters and eventually factory operations in Hefei.

Even amid the financial struggles, NIO shone in China. A 2019 review of its ES6 SUV by the BBC's influential Top Gear called it "very much in the mold of the Audi e-Tron, Mercedes-Benz EQC, Jaguar I-Pace and Tesla Model X."[42] Sales steadily climbed into the tens of thousands, at average prices well north of $50,000.

NIO not only survived its stress test but put itself in position to catch the next wave of investor enthusiasm for EVs that would surface in 2020, following the initial financial scare of the pandemic. On November 5,

2020, the Chinese startup that few Americans had heard of saw its stock market valuation top $53 billion, higher than each of Cheng's old employers, Ford and FCA. Eventually it nudged a nose higher than the 120-year-old icon, General Motors.

By then, Cheng had left NIO for another EV venture. It turned out that the auto show with the Ferrari wasn't the end of his career, just the pivot point. NIO's rise was a new start for him, and a validation. Companies like Tesla and NIO and BYD, which embraced EVs and worked hard to build cool brands, were resonating with the Chinese consumer in a way that the old-school car brands were not. "Toyota, Nissan, Ford, General Motors—they made a lot of money in China, but they just saw China as a joint-venture contract manufacturer. They didn't study China and figure out the user experience," Cheng said.

"I think these traditional guys have a lot of catching up to do."

3

Internal Combustion Busts

O ne summer morning in 2021, Lem Yeung opened an email from Ford's HR department. He stared at his screen. Yeung had worked on internal-combustion-engine development at Ford for thirty years, the exact tenure that made him eligible for the generous retirement package that Detroit's car companies have long offered. The son of Chinese immigrants was also part of Ford power-train lineage—both of his parents also were engineers who worked developing engines at the iconic company.

Yeung's career at the automaker started with an internship and had run the gamut. He had managed the design and launch of a menacingly named V-8 diesel engine—the 6.7-liter Power Stroke Scorpion, the beast used in Ford's award-winning Super Duty pickup. He was involved in many projects spanning the engine spectrum, from a tiny four-banger used in China to the Vulcan 3.0-liter V-6 engine for nineties-era Ranger pickups.

The email was a buyout offer, coming amid one of Ford's seemingly endless belt-tightening cycles. CEO Jim Farley was finally getting some credit for his EV push. But Ford still trailed GM in profitability, and the rookie executive was trying to clamp down on costs. If Yeung agreed to retire early, the company would kick in several more months of salary in addition to retirement pay. For many eligible Ford vets, it might have been a no-brainer. But Yeung was only fifty-two. If he accepted, what would he do for the final decade of his career?[1]

Then again, Yeung's job in the preceding few years had become much less satisfying. For most of his career, the world of engine development at Ford was so big and bustling, Yeung felt like he worked inside the company's nerve center. His first job had been inside a 1920s, wood-paneled building in Dearborn—just a few miles from the historic home of founder Henry Ford—where hundreds of power-train engineers toiled away on generations of Ford engines.

Not far from there were recreational softball diamonds where he played on a team of Ford power-train engineers in the 1990s and early 2000s. Dozens of teams comprised of Ford engineers, divided by skill level, would play one another. But the games got really intense in an all-power-train tournament. Yeung had never sustained a serious injury despite his career as a standout high school wrestler and a judo expert, until while playing shortstop in the softball league, he had his knee blown out by an overzealous engineer trying to break up a double play.

If you think of engineers as quiet, introverted, and immersed in their numbers and calculations, Yeung isn't that. With short-cropped dark hair and an athletic build, he's a bit of an iconoclast, not afraid to spout his opinions or question a supervisor. He tends to sprinkle his conversations with f-bombs. He loved working on engines because of their immense complexity, which demanded improvisation and fostered an entrepreneurial approach to solving problems. He's seen it hundreds of times: eager young engineering grads showing up at Ford, ready to apply their math and computer calculations to their tasks. But engine work, Yeung would explain later, involves far more trial and error and nuance than one might expect. Developing and refining car engines always presents gray areas, judgment calls, and a surprising amount of going with your gut.

Early in his career, Yeung would arrive each morning at a calibration-testing center, where he'd sit in the driver's seat with a computer screen and analyze data including engine speed, ignition spark, and air pressure. He would tweak the inputs over and over and then move onto the next vehicle. Later he would use a spreadsheet to come up with averages, data that would be routed to teams working on new versions of the engine.

"I mean, like, fuck, I don't know how much spark. No one's teaching me. I had to learn later, right? Pressure 10 degrees after top dead center. Where the pre-ignition points are. All that stuff is trial and error. You learn as you go," he said. "I was always worried about bombing out the engine. It was scary. It was super cool."

Ultimately, the decisions made by power-train engineers affect hundreds of thousands of future car buyers, who will, for years, experience the living, breathing engine the engineer created. Yeung remembers being stressed out by the job early on, envisioning all the ways in which he could screw it up. "I was freaking out, because when you get out of school and you've grown up as an engineer and among engineers, you want to be perfect, right? But you realize that's not real engineering."

He was so anxious that, against his better judgment, he confided in his boss, an old-school, Harley-Davidson rider who deployed a sink-or-swim management style. To his surprise, it helped. His boss said: "You know, someone told me this once, and I'm going to pass it on to you, because I remember always being so afraid of fucking things up: 'If you're not breakin' something, you aren't doing nuthin!'" Encouraged by that pep talk, Yeung learned to love his job.

By the time that HR email hit his computer screen, the fun times had faded, Yeung said. New engine programs—the ones that unleashed all that creativity and collaboration among engineers—had dried up. Even though EVs at the time accounted for a tiny sliver of Ford's business, most of the money, resources, and energy was being channeled into developing plug-in models.

Rookie CEO Jim Farley was making bold proclamations about Ford's EV ambitions in an effort to erase the perception that the company had fallen behind. He was hiring engineers from Silicon Valley, from places like Apple and Google—and yes, from Tesla. Traditional power-train engineers, who for so long rode a conveyor belt from midwestern engineering schools like Michigan State and Purdue straight into Ford, were being bought out, let go or marginalized. At Ford, the internal combustion engine was still paying the bills—F-150 pickup trucks, many with growling V-8 engines like Yeung's Power Stroke Scorpion, were generating most of the company's profit. But it was clear that EV and battery specialists—chemists and coders—were the new rock stars.

Since he had been named the CEO in the summer of 2020, Farley had been dropping hints to investors and Wall Street analysts that he the old-guard power-train engineers that Yeung admired so much would be making room for EV specialists.

"ICE [internal-combustion-engine] talent and BEV [battery-electric-vehicle] digital talent are different," Farley told an investor conference in February 2022. "You can't ask ICE people to do certain things. It takes too long."[2]

Yeung initially thought he had an ally in Farley. "I was so excited to have Jim Farley up there. I thought, 'Oh my God, a car guy! A dude who races all the time and reads Jalopnik,'" Yeung recalled, referencing a popular car-enthusiast website. "But man, he disrespected all the engineers so bad. I find it incredibly hurtful."

. . .

The internal-combustion engine, at a basic level, converts fuel and air into mechanical energy, specifically torque, which is what turns the wheels. It starts with fuel and air inside the car's cylinders; smaller cars may have four cylinders; larger or sportier ones might have six or eight. Inside a cylinder, a spark ignites the air-and-fuel mixture, thrusting down a piston inside the cylinder to get things moving.

That symphony of exquisitely timed mini explosions inside the cylinders sets the wheels in motion, but only after an intricate chain reaction of mechanical movements. The pistons thrusting up and down are attached to connecting rods, which attach to a crankshaft. The crankshaft, in turn, uses the energy sent to it and converts it into rotational motion, propelling the wheels. That's the elementary-school explanation; there are myriad other parts that do their piece, from piston rings and flywheels to rocker arms and exhaust manifolds. In modern cars, that whole process is orchestrated by computers—small electronic-control units that help optimize all those movements.

Along the chain, there's a lot than can go wrong. There's also a lot that can be tweaked—incrementally, methodically—to make it better, to eke out just a bit more power, to make things run a bit more smoothly or use a bit less fuel. That's a big part of the appeal for power-train engineers like Yeung: the intricacies of the internal combustion engine allowed for almost endless innovation and problem-solving. It's rewarding work.

The gas engines Yeung worked on have hundreds of moving parts; an electric drivetrain has fewer than twenty-five.[3] An EV consists of a large battery pack, typically with hundreds of rechargeable lithium-ion cells, normally tucked under the cabin floor, sending power to a unit called an inverter, about the size of a small suitcase. The inverter changes the electricity created by the battery from direct current (DC) to alternating current (AC). That power is then fed into an electric motor, or sometimes multiple motors for extra power or to produce all-wheel drive.

The electric motors are also pretty simple, and they pull off a nifty trick that does away with the need for a complex, multigear transmission used in internal-combustion-engine cars. In a gas engine, maximum acceleration or torque is achieved only in a narrow band of operating speed before the engine risks spiking over its revolutions-per-minute limit, which can cause it to fail. That's where a transmission comes in: handing off that torque from first gear to second, or second to third and so on, keeps an optimal balance between all that power coming from the engine and how quickly the wheels are rotating. The engine can deliver faster speeds only as it moves up the gears.

In an EV, an electric motor can essentially take a similar amount of energy as produced by a gas or diesel engine, but use it far more efficiently, eliminating the need for gears. Electric motors deliver maximum torque across all speeds. That's why most people find EVs so fun to drive: that power thrust is there the instant the foot hits the pedal. The acceleration is smooth, without those mini pauses that come from a traditional transmission shifting gears. *Car and Driver*'s 2008 review of Tesla's first car, the Roadster, sums it up this way: "Even more impressive is the instant acceleration at real-world speeds of 30 to 100 mph. Squeeze the throttle, and the Tesla surges forward with effortless ease."[4]

Yeung isn't anti-EV. As a car guy, he's driven plenty of them and knows the rush that all that instant torque produces. His complaint is that, from an engineering standpoint, EVs are, well, boring. They don't require any of that futzing around with air-to-fuel ratios, spark timing, and valve-train tweaking that Yeung and countless other auto engineers have tinkered on for more than a century. Yeung says he and many of his power-train peers could adapt their skills to work on EVs. But that work would involve little of the ingenuity and creativity that he had come to love in the world of gas and diesel engines.

"With an electric motor, there's really not that much to it," he said. "It's no longer an art form."

. . .

As a teenager, Yeung used to hang out with friends in downtown Plymouth—a suburb about twenty-five miles northwest of Detroit—and watch people cruise the bustling main street in Mustangs and Camaros, their rumbling engines echoing off the nearby restaurants and shops. Yeung had to live vicariously through his friends. His parents were too busy working and insisting that he and his brother become engineers, doctors, or lawyers. Still, he was immersed in Detroit's car culture and intrigued by it.

Yeung's father was a physicist and electrical engineer who graduated from Michigan State and ended up in Ford's science lab, where he eventually had a hand in developing the electronic onboard diagnostics (OBD) system, an in-car computer system that monitors and regulates the engine's performance, particularly its emissions. His mother, an electrical engineer, did computer analysis of engines at Ford.

Yeung went to Purdue to get an engineering degree. He graduated in 1991, in the teeth of a recession. He applied for several positions at power companies and was turned down by all of them. Eventually he landed an internship at Ford, and early on, Yeung's work ethic and creativity got him noticed. He caught wind of an employee vehicle-lease program that required managers to fill out a survey when they went to swap cars. Yeung figured that would be a treasure trove of information that could help vehicle developers, but nobody was doing anything with the data. "To me that was crazy," Yeung said. "We have all these engineers filling out detailed surveys about the quality and features of these cars? There's got to be some great shit in there." Yeung went and accessed the data—which at the time, in the early 1990s, required cracking into a mainframe computer—and delivered a thick report of what he found to a vice president.

Yeung remembers one of his favorite engine-development projects, one that underscores the myriad possibilities that power-train engineers have to toy with. Around 2005, Yeung had been looking over Ford's development plan for the 3.0-liter engine that was used in the Escape compact SUV and the midsize Fusion family sedan—two of Ford's top US sellers. Yeung did a double take. No engine upgrades were planned. Normally you'd expect some sort of enhancement every few years to keep those vehicles competitive. But here, for two of Ford's best-selling vehicles, the engines weren't changing.

That was weird, and Yeung couldn't figure out why Ford would do that. Looking back, he suspects it probably had a lot to do with the financial position of Detroit's automakers as Toyota and Honda continued to grow. At the time, though, all Yeung knew was that the auto industry never sleeps; surely the Escape's rivals, like the Toyota RAV4 or the Honda CR-V, would be getting engine upgrades. Marketing in the car business is fueled by whatever new or enhanced vehicle attributes can be touted in multimillion-dollar advertising campaigns. Car brands want to tout more horsepower, better fuel economy, a smoother ride, or a bigger touch screen.

"Are you sure that's a good idea?" Yeung asked his boss of the nonexistent engine plans. "This is going into two of our cornerstone vehicles." His inquiries were mostly ignored. So Yeung decided to freelance an engine upgrade in his spare time.

He recalibrated the engine—fiddling with a higher horsepower by adjusting the valve train, getting more air into the combustion chamber. That required longer, higher-pressure intake runners—a channel that funnels as much of the air-and-fuel mixture into the cylinder as possible. Yeung jury-rigged a mockup of his concept and found it worked, but he would need more than that to move forward. He needed to make sure the changes would not only boost the engine's horsepower, but also eke out a bit more efficiency and maintain a smooth drive. Yeung knew he needed a combustion specialist. Yes, there are engineers whose job is just to figure out the precise air-and-fuel ratio that would deliver the most punch as efficiently as possible. One of the best at Ford was Steve Penkevich, a company lifer and native of Michigan's rural Upper Peninsula, whose title was "combustion technical specialist." Yeung barely knew Penkevich, who was about five years older and had been at Ford longer. When he approached him to gauge interest in the stealth project, Penkevich jumped at the chance.

It turned out that the men shared interests, including hunting and fishing. And Penkevich was every bit the power-train zealot as Yeung—at one point he worked on track simulations for Ford's NASCAR racing engines. A mid-cycle upgrade to an engine that goes into a family car was, on paper, far less exciting. But not to Penkevich. "That's the sort of project where the gasoline in your blood starts pumping," he would say years later.[5]

He remembers the thrill of gathering a ragtag team of power-train engineers into a nondescript conference room to hash over strategies for boosting the Fusion engine's performance. "I think I know exactly how to change that cylinder head if you want to get more power out of it," he recalls saying. "And I think we can do it without making a big tear-up." Power-train strategizing sessions like that, he recalls, were like "watching the *Big Bang Theory*—a lot of like-minded guys all breaking into geek talk, and all trying to solve the same problem."

In the end, the project was barely a blip in the annals of automotive product development: it boosted the Fusion's output by roughly 40 horsepower, to 240 hp, which was still below rival midsize sedans, but at least more competitive. Penkevich was so proud of it, his wife was still driving a Fusion with that engine twelve years later.

Yeung was just happy he finally got buy-in for his under-the-radar project. "Sweet, I didn't get fired," he remembers thinking. He and Penkevich became fast friends and still golf and fish together, even after Penkevich caught him with a hook while the two were fly-fishing on the Huron River.

. . .

When Yeung got the buyout email from HR, electric vehicles were still barely noticeable on the US sales automotive sales charts. At GM, Volkswagen, Ford, and other big car companies, EVs accounted for less than 2 percent of sales—and they were money losers. Overall, just 2 percent of all US new-car sales were fully electric models, and Tesla accounted for roughly three-quarters of those. To the outsider, the electric-car revolution seemed like an abstract, faraway idea.

And yet, inside the car companies, the hard pivot to electrics was in full swing and upending a century of internal order. A generation of engineers like Yeung were feeling pressure. Thousands of mom-and-pop suppliers who for decades had made steady profits from selling a few specialized widgets that go into gas-engine vehicles were trying to remake themselves, aware that their future customers' gas-engine orders—while likely to continue for many years—were going to steadily shrink.

For more than forty years, tiny Michigan auto supplier P.J. Wallbank Springs has cranked out essentially one part: a circular metal pack of small coil springs used in automatic transmissions. The company—located near the shores of Lake Huron, a few miles from the Canadian border—is really good at springs: its component finds its way into roughly 10 million new vehicles annually, more than 10 percent of all cars sold on the planet. But electric cars don't have multi-gear transmissions.

That existential question isn't lost on the chief executive, Chris Wallbank, grandson of founder Philip John Wallbank. In 2020, he created a new business that works with startup companies to build components for electric and autonomous vehicles. "We don't want to be left making the best buggy whips," he said.[6]

Even if it succeeds, some suppliers won't. There simply aren't enough parts to source. Carmakers have been redirecting the money they spend to develop new gas-engine vehicles, engaged in an all-out arms race for EV spending and bragging about it. In 2018, Ford executives boasted that they would dole out $11 billion over the next few years to develop electric cars. In 2020, GM said it would invest $20 billion in EVs, along with driverless tech. Ford a year later said $30 billion. GM came back after that and pegged $35 billion. Ford in early 2022 upped the ante yet again, to $50 billion.

Developing a new engine program can cost hundreds of millions of dollars, even north of $1 billion, and involve hundreds of suppliers. In 2011, car companies globally rolled out almost seventy new engine families—a base engine that can be used as the foundation for several variants, according to research firm S&P Global Mobility. By 2018, the number of new engine families was twenty. In 2021, it was five. The research firm expected the number to reach zero this decade, meaning virtually no engines are being designed from scratch.[7] Sure, gas and diesel engines will be in showrooms for a few decades to come. But the expectation is that car companies will stop sinking money into the machine that they'd spent a century working to perfect.

The industry's EV push is reversing a collective decision made more than a century ago, when electrics competed with internal-combustion vehicles and even steam-powered automobiles to become the power train of choice for the car business. Around the turn of the twentieth century, cars powered by electric motors outnumbered those running on gasoline in the United States. It was like a higher-stakes version of Betamax versus VHS videotape.

An article from the July 1900 issue of the *Automobile Magazine* delved into the contradictions of electric cars in a way that's eerily contemporary: "In cleanliness, ease of management and safety, it has ideal qualities," it read. "Its great cost, its excessive weight and the various shortcomings of the storage battery, together with its extremely limited radius, combine to limit its field to certain forms of urban use."

By the early 1900s, the reliability and quality of gas engines had improved with innovations like the electric starter and ignition system, both invented by engineer Charles Kettering, who became the longtime head of research at GM. And though gas stations were just emerging, hardware stores sold cans of petroleum. Meanwhile, electric generation outside big cities was scarce. Motorists with electric cars then had range anxiety, the fear of being stranded on the side of the road with a bricked battery, far from any means to recharge.

Gasoline engines won out. Henry Ford's assembly-line innovations dramatically drove down the cost of vehicle assembly. That made his mass-produced gas-engine Model Ts far cheaper than the electric cars sold by dozens of upstart companies. As road systems improved, Americans wanted to travel longer ranges, which also favored gas cars.

In the ensuing decades, generations of power-train engineers from Detroit to Tokyo and Stuttgart, Germany, toiled to refine the machine that powered global transportation. Stepwise changes were rare; through mostly incremental tweaks and tricks, like Yeung's modest improvement in his rogue project, they made the internal-combustion engine more powerful, smoother, more responsive, more compact—and critically, more fuel efficient. In 1966, the Ford Mustang was considered a monster, and it had an option for a V-8 engine that could generate 210 horsepower and go from a standstill to 60 miles per hour in 8.2 seconds.[8] The 2023 Mustang has a similar-sized engine that makes 470 horsepower and screams to 60 miles per hour in about 4.4 seconds.

Over the decades, attributes like engine size, shape, and even how an engine was cooled—air or water—became grist for legendary debates among car enthusiasts. Names like Jaguar's straight-six engine and Chevrolet's small-block V-8 denoted the size and layout of the engine's cylinders, but also became powerful brands unto themselves. Car companies built entire marketing campaigns around their engines, with names like the Cadillac Northstar or Dodge Hemi.

The power train was once so central to a car brand's identity that, for decades, General Motors operated separate engine-development divisions for most of its brands. The idea was that, if the engine was the heart of the automobile, then each brand should have its own bespoke motors, at least for larger, more expensive cars. In 1965, GM offered more than twenty-five different gas engines across its Chevrolet, Pontiac, Oldsmobile, and Buick brands, nearly all developed, tested, and manufactured separately from one another.[9]

Eventually, even GM—for much of the twentieth century the world's largest and most profitable corporation—came around to the cost-cutting revelation that its brands could share components or even entire engines without vehicle owners noticing. By the mid-1970s, GM had quietly begun joint engine development among its brands, sharing parts and in some cases using essentially the same engine in cars from different marques. That saved money, but back then, the notion of the engine being central to the car's identity was so ingrained in consumers' minds that the discovery of GM's move boiled over into a PR crisis.

In 1977, a Chicago man named Joseph Siwek brought his Oldsmobile Delta 88 big sedan in for engine work. His mechanic ordered a few replacement parts, but when they came in, he discovered that some of them didn't fit. Eventually it was determined that the 350-cubic-inch-displacement V-8 engine in his Oldsmobile—a brand that was above Chevy in the GM pecking order—used a down-market Chevy engine of the same size. The man sued GM for false advertising, triggering national news coverage of GM's "engine scandal" and a multiyear legal headache.[10] Eventually, a federal jury awarded some Oldsmobile owners $550 refunds, costing GM about $8 million.[11]

That didn't stop the consolidation though. The twenty-five different engines GM used in 1965 are gone. In the mid-2020s, the company uses fewer than a dozen different gas and diesel engines for all its US cars today.

· · ·

Yeung was just one soldier in an army of power-train engineers whose collective problem-solving abilities and earnest sleuthing helped evolve the internal-combustion engine and solve some of the industry's most vexing engineering challenges.

The federal Clean Air Act in 1970 required car companies to reduce the pollution that their tailpipes were spewing. Catalytic converters were added to reduce toxic fumes. New injection systems were developed to spray fuel directly into the engine's combustion chamber, boosting power and efficiency. Companies added more gears to transmissions to help engines run more optimally, also saving fuel. Eventually, hybrid gas-electric cars, popularized by the Toyota Prius, deployed a small battery pack and electric motor to significantly enhance fuel economy.

For the 2022 model year, US vehicles averaged 26.4 miles per gallon—roughly double from 13.1 mpg for 1975 models, according to Environmental Protection Agency data.[12] The amount of greenhouse-gas-producing carbon dioxide that US vehicles belched out was roughly cut in half per car over that period, the EPA says, much of it thanks to power-train engineers.

Over the years, engineers were forced to get more creative to achieve even small efficiency gains. Starting in the early 2000s, one curious advancement became widespread across gas-powered cars: stop-start systems, which cut the engine when the driver idles at a stop, firing back up when the foot is lifted from the brake. The awkward feature—which is estimated to bring about 3–5 percent in fuel savings—has befuddled its fair share of car owners.[13]

By the late 2010s, car executives were facing down tightening regulations globally. Carbon-emissions rules in Europe had gotten so strict that some automakers were being threatened with fines of $100 million or more if they couldn't improve their emissions.[14] Auto execs were concluding that they would never be able to wring enough efficiencies from the internal combustion engine to meet the rules in the long run.

At Ford, that sentiment wouldn't have come as any surprise to Yeung. He had seen all the new engine programs dry up. It made him uneasy, not just about his own future, but about the future of Ford, the company that had sustained him and his family for more than fifty years.

In Yeung's view, the ability to engineer, refine, and produce gas and diesel power trains was the secret sauce of the auto industry for a century. It was so complex, you had to be among the biggest corporations on the planet, with the deepest of pockets, to do it at scale. It gave the big car companies huge barriers to entry that allowed them to operate for decades with virtually no newcomers. Of course, there were regional rivals that could crash into a given market and wreak havoc—notably, the US invasion by the Japanese automakers starting in the 1970s, which ate into Detroit's market share. But those were traditional automakers breaking into new markets. There essentially were no startups of note, until Tesla, a hundred years into the car business, when it started selling refined electric sedans with head-turning designs direct to consumers.

Now, as Ford and other automakers pivot to electrics, they are competing with big players from halfway around the world. Companies like Samsung and LG and Panasonic are big dogs on batteries. Electric motors also are largely the domain of Asian players big and small, including Japan's Nidec Corp. and many low-cost options in China. And the stuff is relatively easy to put together into a vehicle package. Some car execs have compared it to snapping a Lego set together.

Yeung is not convinced the car companies can establish enough expertise in EVs to stay on top. They're losing their competitive advantage, their expertise. They're losing guys like Lem Yeung and Steve Penkevich. The built-in advantages that the incumbent carmakers wielded for more than a century are disappearing as cars become less mechanical and more infused with complex software. They are entering a race in which they're starting from behind, with companies that suffer none of their legacy drag. Tesla and Chinese EV maker BYD—and a slew of other Chinese EV upstarts—are outmaneuvering the GMs and Volkswagens on lower-cost battery setups and consumer-friendly tech features. Thinking like a car company has flipped to become a disadvantage.

"EV technology is not that difficult. It's just modular," Yeung says. "Ford can learn to build and house their batteries and their motors, but honestly, the innovation level on a motor and the ability to do it uniquely versus an engine? They won't be able to do it cheaply enough. I don't think the auto industry is going to innovate in battery technology or electric motors. I think all the barriers to entry have gone."

Yeung had come to these conclusions long before that HR email landed atop his inbox. His buddy Steve Penkevich had concluded the same, having taken early retirement eight months earlier, in December 2020, at age fifty-nine.

Yeung clicked accept.

4

Hippie Power

n 2016, it was still more than a year before GM execs would start talking about an all-electric future. Rivian was still a clandestine operation. Tesla was just a startup, still a few years away from its Model 3 breakthrough. But Volkswagen's design studio was bustling more than usual. At the crisply lit, white-walled building at company headquarters in Wolfsburg, Germany, designers tried to keep pace with a crush of electric-vehicle programs the company was pinning its future on.

VW engineers had developed a new EV platform that would underpin more than a dozen electric models. A large battery pack would be tucked under the floor, freeing space in the cabin. The removal of an engine block up front would give designers a blank canvas to style new front ends.

Felipe Montoya was feeling at once overwhelmed and pumped. The Brazilian-born designer who favors trim sport coats with jeans and sneakers, and has a shock of jet-black hair, had made a name for himself crafting the exterior of the latest generation of the VW Golf, the sporty hatchback, one of the company's top sellers. Montoya had never worked on an electric car. But, as he and his colleagues soon realized, VW's electric onslaught would free them like never before. His sketch pad was full of ideas.

And tucked deep inside that sketch pad were secret renderings of a vehicle that hadn't yet been given the official go-ahead: an electric version of the iconic VW microbus. There was heavy speculation it would become reality, probably because the idea was almost too perfect: the VW bus—a symbol of 1960s flower-power hippie culture—reincarnated as an earth-friendly electric. Montoya and other designers wanted to be ready when-ever top brass gave the green light.

Volkswagen first introduced the toaster-shaped bus in 1950. Named the Type 2, its split windscreen, snub nose, and optional two-tone paint job gave it a cute, disarming appeal.[1] In America, the bus emerged in the 1960s as a vehicle for families to transport camping gear, and for young people to mobilize for concerts and antiwar protests. It was common for long-haired owners to paint flowery murals over the exterior and swap out the VW badge for a peace symbol. Montoya remembers seeing VW buses plying the streets of Brazil where he grew up.

His fixation on designing an electric version of the van wasn't rooted in nostalgia. Montoya and other designers knew that this new EV platform was a chance to return the bus to its roots, design-wise.

The original Type 2 had its engine mounted in the rear of the vehicle and, therefore, had no need for a long hood to house an engine block. That allowed for the driver's seat to be pushed way forward and gave the bus its distinctive silhouette, with its flattened front end. It also made room for the bus's massive "greenhouse"—designer speak for the windows—with inviting rectangular openings that owners invariably would cover with Grateful Dead dancing-bear curtains or trippy tapestries.

But then, later iterations moved the engine to the front, which needed more space and gave the bus a nose. For decades, the VW bus became just any other commercial, boring van on the road. That distinctive profile revered as a symbol of the counterculture sixties became a hazy memory.

"With this new electric platform, we knew that we'd have the chance to finally bring back that silhouette," Montoya said. "It was the chance of a lifetime."[2]

. . .

Of all the big legacy automakers, VW made the earliest and deepest bet on electric cars. It was born of desperation and scandal.

In 2014, a group of researchers from West Virginia University flew to California and rented three diesel cars to do some on-road testing. The team wanted to see if these vehicles—VW Jetta and Passat wagons and a BMW X5 SUV—were meeting the EPA's emissions rules. The engineers stuffed the hatchbacks with testing equipment the size of small suitcases, connected to a nest of tubes, and a probe inserted into the tailpipe to analyze the cars' emissions. They put the diesels through their paces in different settings along the West Coast—from the hilly outskirts of San Diego to the jammed highways around Los Angeles, the steep, congested streets of San Francisco, and wide-open Interstate 5 up to Seattle.[3]

Diesels generally are used in big, commercial trucks. They deliver a lot of low-end torque—that initial jolt that gets a vehicle going from a stop. And, importantly, they return better fuel economy in highway driving compared with gasoline vehicles. The downside: diesels spew a disproportionate amount of nitrogen oxides, or NOX, a highly reactive gas that contributes to acid rain, hazy air pollution, and water contamination. In humans, it can lead to asthma and other respiratory problems.

Diesels were the power train of choice in Europe and had been for a few decades, partly because governments taxed them at lower rates than petrol cars. Regulators liked their fuel efficiency and saw diesel tax breaks as a way to benefit commercial truck drivers. Thus, in Europe, diesel accounted for 60 percent of passenger-car sales in 2011.[4] But around that time, European policy makers began questioning whether the trade-off—better fuel economy but more-noxious emissions—was sound policy. The European Union was tightening emissions rules for both gas and diesel cars.

Volkswagen was a dominant force in diesel engines and one of the world's biggest carmakers, usually neck and neck with Toyota for the global sales crown. The company was founded in 1937 by the German government, then under the control of Adolf Hitler's Nazi Party. Hitler wanted an affordable car for the masses—a "people's car," translated as *volkswagen*. He appointed renowned automotive engineer Ferdinand Porsche to develop a round-shaped car that would eventually assume the name "Beetle." As Germany recovered from World War II, VW would crank up Beetle production worldwide. Sales lagged in the United States at first because of the car's Nazi roots, but it eventually caught on. The Beetle would go on to break a sales record held by Ford's Model T, selling more than 15 million copies worldwide.[5]

VW's diesel cars helped the company become a dominant player in virtually every major regional auto market—Europe, South America, China—except North America. When VW introduced diesels in the United States in the 1970s, it was a novelty; few American car buyers were interested in diesel-burning small cars. Eventually VW did OK with diesel versions of the Rabbit, a small, boxy hatch. But overall, VW's sales in the United States bobbed along for decades, far behind the Japanese automakers that were vacuuming up US market share.

VW's status as a bit player in the American auto market stood in the way of its broader goal of being the dominant number one in vehicle sales globally, finally propelling itself clear of Toyota. By the mid-aughts, VW executives thought a reimagined diesel engine could be the edge they needed to gain ground in the US market. The company planned to introduce its so-called clean-diesel system from Europe, which executives said dramatically reduced nitrous oxide and thus could be sold in all fifty US states, even in emissions-strict California and New York. VW drummed up publicity with a four-month, cross-country "Dieselution Tour." A fifty-three-foot-long semi-trailer traveled state to state with displays touting VW's environmental cred and the 35 percent bump in fuel savings that the forthcoming diesel models would get.[6]

The timing was fortuitous: its first clean diesels—marketed under the name TDI—reached showrooms in 2008, just as prices at the pump were soaring above $4 for the first time. The company's marketing trumpeted the remarkable combination of fun-to-drive, full-of-torque diesels with fuel economy comparable to hybrids; some owners boasted returns of more than fifty miles per gallon. VW's modest US market share doubled within a few years.

But something seemed off—at least to researchers at a small nonprofit called the International Council on Clean Transportation. The group had seen reports out of Europe showing that road tests for diesel cars detected emissions that often blew past the official limits. The ICCT hired the West Virginia scientists to conduct those road tests out west.

The lab-rat engineers were shocked by what they found on their jaunts around the Pacific coast. The VW cars were failing to stay within the federally mandated limits for NOX. And not by just a little. The Jetta spewed as much as thirty-five times the amount of the pollutants allowed into the air, and the Passat as much as twenty times. The BMW, on the other hand, passed the emissions tests.

"We were astounded when we saw the numbers," John German, an ICCT senior fellow involved in the testing would say later.[7]

In May 2014, the group issued a 117-page report, innocuously titled "In-Use Emissions Testing of Light-Duty Diesel Vehicles in the United States." Its takeaway, in the most objective, scientist-speak manner possible, was that VW was cheating. The study landed with little notice. But regulators at the EPA and in California quickly and quietly opened investigations. After much back-and-forth with VW officials in the United States, the company told regulators it had figured out the problem and issued a recall—a software patch to bring the vehicles into compliance. But regulators tested some cars after that fix. They still were spitting out way too much pollution.

On September 18, 2015, the EPA released a six-page statement accusing VW of purposely evading the Clean Air Act. The company had installed "defeat devices"—software with advanced algorithms that could detect when the vehicle was being tested for its emissions levels. Under testing, the pollutants streaming from the tailpipe were kept in check. In real-world driving, though, they spit out far more particulates. The company was cheating to get around US environmental rules, on a massive scale. About a half-million cars on American roads had the defeat devices installed. Globally the number was 11 million, mostly in Europe. These regulators had just blown the lid off one of the biggest corporate scandals of the twenty-first century.

The fallout was swift and fierce. Sales of TDI Golfs and Jettas were halted immediately. US politicians and environmental groups eviscerated VW. Then-CEO Martin Winterkorn resigned within days. Investigations were opened by an alphabet's soup of federal agencies—the DOJ, SEC, FTC, and more. Within weeks, VW execs were dragged in front of congressional hearings. More than a dozen VW officials and employees would be charged with crimes.[8] The company eventually pleaded guilty to criminal conspiracy and other charges. VW was forced to either buy back customers' cars or pay them thousands of dollars in restitution. Those costs, along with government fines and other settlements related to the scandal, would hit a staggering $35 billion—roughly three years' worth of profits.[9] VW's reputation was shredded.

Nearly a month after the crime came to light, a group of about ten top VW executives huddled during a weekend retreat at a genteel inn, tucked in the forest near Wolfsburg. A Saturday afternoon meeting was led by Herbert Diess, who had joined the company just two months earlier to become head of the VW brand, or what was left of it, after a long career at BMW. Diess was well-spoken, fluent in English, and a green-car zealot; at BMW, he led the launches of two landmark electric cars. The discussion centered on the damage the scandal had wrought on the VW name and ways it could be resurrected—conversations that culminated in a massive strategic shift to electrics.

"It was an intense discussion," Jürgen Stackmann, VW brand's board member for sales, said afterward. "So was the realization that this could be an opportunity, if we jump far enough."[10]

The following week, VW announced it was rolling out a brand-new vehicle platform, dubbed MEB, to serve as the building blocks for an arsenal of EVs of various sizes and styles. In the coming months, the rest of the industry would discover just how deep VW's redemption gamble would go: the automaker was earmarking $40 billion to primarily develop and build EVs through 2022, a sum that would grow sharply in subsequent years. VW leaders were all in: the company aimed to surpass Tesla to become the world's top EV seller by the mid-2020s.[11]

It was a remarkable strategic pivot, one that would have major implications for the rest of the industry. No traditional automaker at that point was talking about building EVs at this scale. To the legacy car industry, EVs still were essentially a science experiment. Most carmakers tolerated selling small numbers of money-losing EVs to meet regulations so they could continue to sell their profitable gas guzzlers.

But the VW executives knew that, even though diesels commanded more than half the new-car market in Europe, the scandal had accelerated the demise of that time-tested power train it had built the company on. That one of the world's top two carmakers would essentially bet its future on electrics was a seminal event for the rest of the industry.

Back in Wolfsburg, designers gathered in the studio amid the glow from oceans of digital screens. On display was the future lineup of VW electrics—the so-called ID family of battery-powered models. ID.3, a small SUV, was to be the first to market, within three or four years. There were at least a half-dozen others. The showstopper, though, was the reincarnation of the bus, which would take the name ID Buzz, a cross between the sound of electricity and the "bus." There were massive, printed images of every generation of the bus, going back to the famed T2 from the 1950s.

Montoya and his colleagues had pulled out the sketches they'd been noodling in private. It was clear from the start, though, that the design for the battery-powered microbus couldn't be a radical rethink of the original. It had to have just enough of a modern touch and also had to resemble the rest of VW's forthcoming ID electric lineup. And yet, there was no sense in bringing back the beloved hippie mobile unless it was instantly recognizable.

"We always had the T2 as a target," Montoya said. "We just had to put all these different ingredients together, shake the drink, and see how it would taste in the end."[12]

. . .

On a frigid January morning in 2017, hundreds of journalists packed around the VW stand at the Detroit auto show. In a promo video, one executive promised that the vehicle about to appear on stage was "the reinvention of the soul of the brand." Moments later it rolled out: a bright yellow-and-white van with a digitally lit VW logo gleaming on its face. Diess followed the van onto the stage. Tall and slender, in his late fifties, he wore a trim navy blue suit and open-collar white shirt, and stood beaming in front of the ID Buzz. The CEO right away addressed the elephant in the room, proudly declaring that the company had moved past the challenging Dieselgate scandal. VW, he explained, was embarking on the biggest transformation in its history.

"We will inspire America with great electric cars," Diess said. The company would also invest about $2 billion to install EV chargers across the United States, he proudly declared, sidestepping mention of the fact that VW was required to do so under terms of its Dieselgate settlement with federal regulators.[13]

The boxy van had a driving range of 270 miles and slick features like swivel seats. It was officially a concept car—a prototype that carmakers show publicly as a sort of fun trial balloon. VW executives strongly hinted that the bus would get the nod to go into production eventually, though. Not everyone was convinced, given the industry's history of flirting with cool EV concepts that never saw the light of day.

"It's a brilliant marketing strategy," the Verge tech website wrote. "But the question remains: will Volkswagen ever build this thing?"[14]

VW's electric turn sharpened a year later, when Diess was elevated to become VW's CEO. The company's supervisory board liked how Diess had navigated VW through the Dieselgate crisis as brand chief. It also didn't hurt that he had boosted the division's profitability. Diess was still a relative outsider, and the board wanted change from the stultified, insular culture that produced the cheating scandal. And they thought the forward-thinking Diess was the guy to prepare VW for battle with Tesla, Uber, and other tech players out to reshape the world of mobility.

Self-confident and even-keeled, the mechanical engineer and manufacturing expert—with a PhD in "assembly automation"—saw himself as a change agent, prepared to lead VW across the risky chasm from old-world, gas-guzzling analog cars to electric, digitally connected ones. He was known to take commercial flights, rather than the corporate jet flown by other VW execs, and kept a neat, spartan office in Wolfsburg. He wasn't a monk, though: Diess drove a Ferrari with personalized plates and co-owned a trendy Spanish tapas restaurant in Munich.[15]

Around this time, in 2018, more car executives had started looking over their shoulders at Tesla with a greater sense of fear and respect. Diess sent their angst to the next level. VW planned more than a dozen factories worldwide to make electric cars, including many in Germany. The company envisioned some *seventy* electric models over the next decade. Diess initiated plans for VW to make its own batteries. And he was gunning for Tesla's EV sales crown.

"It's an open race," Diess said. "We are quite optimistic that we still can keep the pace with Tesla and also at some stage probably overtake them."[16]

But he wasn't trash-talking Elon Musk's company, as some previous VW execs had. In fact, he lauded the brash billionaire as the industry's standard bearer on electric cars. At times, he almost seemed obsessed with Musk and Tesla. And Musk, in turn, was keen to return the flattery.

"Herbert Diess is doing more than any big carmaker to go electric," Musk tweeted in 2019. "The good of the world should come first. For what it's worth, he has my support."

Weeks later, Diess defended Tesla when reporters, during a press event in Wolfsburg, questioned whether Musk's company was big enough to be a viable threat to VW and other established carmakers. "Tesla is not niche," Diess said. "It's a competitor we take very seriously." Their bromance transcended social-media niceties.[17] The two were said to have spoken on the phone several times. In 2020, Musk visited with Diess in Germany and even got a test drive of the ID.3.[18]

Still, Diess became deeply worried that Musk's company, which at the time was selling about one-twentieth the number of cars as VW globally, had gotten too far ahead on EV know-how and software expertise. In a leaked internal memo, Diess conceded that the Tesla threat was giving him "headaches." He worried that the sophistication of Tesla's central computer—which was capable of delivering near-constant updates to features like automated steering, and siphoning data to constantly improve the capabilities—was pulling away.[19]

In an if-you-can't-beat-em-join-em moment, Diess invited Musk to speak via video at a special meeting of about two hundred VW managers in Austria, in the fall of 2021. Diess, dressed in jeans and a wool sport coat, sat on stage next to a giant screen of Musk's beaming face. He asked, essentially, how Tesla had done it. Musk said something about his management style and his understanding of engineering and supply chains. It was a surreal scene: the CEO of a proud, global, behemoth automaker asking this pesky entrepreneur who had been casually dismissed a couple of dozen months before how he'd been able to outmaneuver this entire, massive auto industry.

Diess's worries mounted. Tesla was months away from opening its first European plant—just a few hours' drive from Wolfsburg, near Berlin. The Tesla factory was expected to eventually crank out a half-million Model 3s and Ys a year. The VW CEO was hearing that Tesla would be able to build its vehicles roughly three times faster than it took VW at its main electric-vehicle factory.

In November 2021, Diess articulated his growing angst during an address to thousands of workers at VW's biggest factory—the world's largest vehicle-manufacturing complex—in Wolfsburg. The first buildings on the campus went up in 1938. More than 20,000 people work on the plant grounds, screwing together Golfs and Tiguan SUVs. Another 38,000 work at the nearby corporate headquarters. Hard on the banks of the Mittelland Canal—picturesque, in an industrial sort of way—the complex includes an iconic power plant, with four soaring, brick smokestacks.

"I'm worried about Wolfsburg," Diess told the crowd. It was the first big employee meeting since the onset of the Covid-19 pandemic.[20] EVs had taken off in the two years since they had last gathered. More precisely, Tesla had taken off. The Model 3 would become the number one vehicle in Europe. And startup EV makers were chewing into VW's market share there. These newcomers were beating VW at its own game. They had figured out how to crank out battery-powered cars at scale, profitably, in a way that the eighty-three-year-old automaker had not.

"I want that your children and grandchildren can still have a secure job here with us in Wolfsburg," Diess said. "That's why I'm here."

Just weeks earlier, Diess had been even more frank about the prospect of major job losses stemming from the EV transition, because electric cars don't require as much manpower to assemble. Word leaked from a private meeting with VW's supervisory board: Diess had warned that thirty thousand VW jobs in Wolfsburg could be in jeopardy if the company didn't accelerate its transition to electric cars to better compete with Tesla and other interlopers.[21]

That sober forecast surprised and rattled board members, half of whom—under VW's labor-centric governance structure—were representatives from Germany's IG Metall trade union. Diess had clashed with VW's labor leaders before, as his cost-cutting focus triggered consternation about job losses. There was growing weariness of Diess's blunt messaging, which they might characterize as doomsaying.

It didn't help that Diess was stumbling on his signature strategy: using VW's manufacturing heft to bring lower-cost EVs to the masses. VW was struggling to roll out the first of all those electric models, the ID.3, because of software glitches. It was a bad look for VW and other legacy carmakers that also were botching their early EV launches—the old metal-benders couldn't figure out the technology as cars were becoming more digital and powered by electrons.

The labor faction on VW's board smelled blood in the water and were pressing for a no-confidence vote that could force Diess out. They'd had enough of this change agent and his Tesla fixation. Was it because the truth was hard to hear?

Diess was defiant: "I'm being frequently asked why I keep comparing us with Tesla. I know this is annoying to some," he said in November 2021.

"Even if I no longer talk about Elon Musk," Diess said, "he will still be there."[22]

5

Charging Up the Heartland

I n the fall of 2019, around the same time Elon Musk was lauding VW's efforts on electric cars, David Jankowsky was driving his blue Toyota Avalon hybrid down rutted-out gravel roads of eastern Oklahoma. He pulled into Wagoner, a town of about eight thousand on the edge of the Cherokee Nation tribal area. A forty-five-minute drive southeast from Tulsa, Wagoner is surrounded by barren fields, farmland, and ranches. An annual bluegrass and chili festival draws thousands each fall to Wagoner's faded downtown, home to a shoe-repair shop, a bail bond office, and Smokin' Sisters BBQ restaurant. A chain of large lakes on Wagoner's eastern edge, created by dam projects in the mid-twentieth century, offer relief from the unbroken Plains, attracting retirees and anglers.

A Georgetown law graduate and entrepreneur, Jankowsky was scouting the town as a potential location to install electric-vehicle charging stations. The fact there might hardly be anyone for a few hundred miles in any direction with an electric car in their garage didn't seem to bother him.

Jankowsky was determined to make a living in the green economy and was convinced that EVs—then barely noticeable on US roadways—would one day find their way even to lonely stretches of the Plains like this one. After a solar-and-wind energy firm he had worked for went belly up, he started Francis Energy to get ahead of the demand.[1]

Francis Energy doesn't install the garden-variety chargers found in garages and outside shopping malls. It deploys high-speed chargers, capable of 350 kilowatt-hours of power. They can charge a larger, seven-passenger SUV—from nearly depleted to an 80 percent charge in about a half hour. But these machines are expensive, ranging from around $50,000 to north of $150,000. Jankowsky's gamble was risky. He had to sink lots of capital into building something based on a bet that the users will come.

Still, his conviction was so strong that he not only made the bet but was placing it in the unlikeliest of places. In Oklahoma, just about everything works against the idea of electric vehicles. The state's biggest industry is oil and gas. Brawny, gas-thirsty pickups are popular, and many vehicle owners drive longer distances than typical commuters. Its politics are deep red. Start a conversation with an Oklahoman about charging amps or kilowatt-hours, and you're likely to draw blank stares or maybe scorn. In 2023, it ranked forty-eighth among states in EV registrations, at just 1.8 percent of new vehicles sold.[2]

In his mid-forties, with salt-and-pepper hair, lean frame, and a radio voice, Jankowsky grew up around Washington, DC, but has Oklahoma roots. His grandfather was an influential Democrat in state politics in the 1960s. Jankowsky's family moved to Washington when he was young because his father, attorney Joel Jankowsky, got a job as an aide to the Speaker of the House and Oklahoman Carl Albert.[3] Jankowsky followed in his father's footsteps by getting his own law degree. He worked as a lawyer for several years until 2010, when he took a job with SunEdison, then a small renewable-energy developer poised for big growth. As a vice president, Jankowsky was based in Singapore, trying to expand the company's reach into Asia. It was exhilarating work, and for a while the company's stock soared. When the company eventually hit financial straits, Jankowsky bailed, just before its 2016 bankruptcy.

He headed back to Oklahoma for the first time since he was a boy. "As an entrepreneur, you're looking for problems to solve. And this area, in the middle of the country, has a lot of them," he would say later. His initial startup was a solar company, leveraging his past experience. But it struggled, so Jankowsky pivoted, inspired in part by Tesla, which was hitting its growth spurt, and by Mary Barra and other auto execs publicly proclaiming they were all in on EVs.

He knew that EV charging in the heartland was going to be hard to build a business around, but he went for it anyway. Jankowsky put his lawyer hat on and began poring over the Oklahoma tax code, searching for any government money available for EV charging. It turns out that Oklahoma—while in many ways a terrible place to bet on the EV transition—was in this one respect perhaps the best place in the country.

Tucked deep inside the state's Alternative Fueling Infrastructure Tax Credit statute was a potential windfall for Francis Energy: the state would cover up to 75 percent of the total project cost of an EV charger installation.[4] If a station with six fast chargers cost $1 million to install, the property owner would qualify for a $750,000 tax credit.

That level of subsidy was a game changer to convince otherwise reluctant hotel operators, rest-stop operators, and other property owners to install a few chargers, Jankowsky figured. It was a long-term bet that EVs eventually would emerge as the go-to power train in the broader transportation industry. In Oklahoma, specifically, he had conviction that electric pickup trucks would supplant the gas guzzlers now rumbling down those dusty roads. And if Francis Energy had locked up the highways and interstates, all those plug-in trucks and their giant batteries would be sucking power out of his company's charging stations.

The tax credit could be the difference. But there was a time crunch. On December 31, 2019, the subsidy would be cut nearly in half. Most of Francis Energy's twenty-person staff had fanned out that fall, hitting town halls, hotels, and roadside diners to pitch property owners with a hard-to-pass-up proposition. Between the tax credit and Francis Energy's backing, they wouldn't have to spend a dime to have a state-of-the-art fast charger installed. In some cases, depending on how their tax bill shook out, the property owner could make some extra money.

That vision was just a mirage in the scrubby landscape when Jankowsky pulled into Wagoner and found his way to the low-slung, stone-sided town hall. Jankowsky walked into the municipal building and asked to see

Mayor Albert Jones. The town engineer was in his office at the time. He told Jankowsky the mayor was out to lunch.

"Where is he? Is it OK if I stop by?" Jankowsky asked.

The engineer walked him to a nearby deli for an impromptu meeting with the town's top man, who went by Mayor AJ. It's fair to say that Mayor AJ had never given a second thought to electric-car chargers before. That a Georgetown-Law attorney had driven hours just to chat with him about EV charging over lunch seemed even stranger still.

"Mr. Mayor, I'm going to put EV chargers right over there, in your parking lot," Jankowsky said, pointing across the street to the town hall. "It's going to cost you nothing. And by the way, I'm going to hook them into the utility."

Jankowsky knew that Wagoner was one of the few towns that provided utility services to its community. He was not only offering the mayor free $100,000 chargers, but also offering to bring the customers right to his front door. "We're going to be buying a lot of power from you," Jankowsky said.

Mayor AJ looked up across the table from his sandwich. "Tell me where to sign."[5]

. . .

Jankowsky is among a slew of entrepreneurs who see a business opportunity in helping to stitch together America's skimpy, hodgepodge EV-charging network. The United States by some measures is a decade behind on the charging infrastructure needed to sustain the sharp increase in EV sales the industry envisions. It also is way behind China and Europe.

China's EV public charging infrastructure was more than ten times the size of North America's in the early 2020s. Europe's was about four times the size.[6] The sheer speed at which China is lighting up the country with charging stations is staggering. It installed nearly 600,000 in a twelve-month span ending in late 2022—about 1,600 new charging stations went live every day for a year.[7] That's more than the United States aims to install by the end of the decade.

By one count, a single province in southern China, Guangdong, as of late 2021 had 350,000 public chargers, three times more than the entire United States.[8] Guangdong is about the size of Oklahoma. "Range anxiety" in the coastal economic powerhouse that borders Hong Kong "is a thing of the past," a Bloomberg article said.[9]

Of course, it could be that Europe and China will need more public EV chargers than the United States in the long run, as both regions have greater population densities, which means higher concentrations of people living in apartments and other multiunit buildings. That reduces EV owners' ability to charge at home.

Concern about a lack of EV chargers ranks as one of the top reasons why most Americans aren't ready for EVs (the other one: high prices relative to gas cars). In 2023, a J.D. Power survey of more than eight thousand Americans turned up some good news for EV makers: about six in ten respondents were at least somewhat interested in buying an electric as their next vehicle. For those uninterested, a lack of places to charge was the biggest deal breaker.[10]

There are three different types of EV chargers. So-called Level 1 chargers plug into a standard, 120-volt outlet found in most homes. It works, but the electrons trickle into the car battery at an excruciatingly slow rate. The roughly 1.2 kilowatts per hour of electricity it delivers can take longer than twenty-four hours to fully charge the vehicle—even multiple days for a big battery.

That's why most people who buy an EV and have a private garage opt to install a Level 2 charger. These deliver at least 7 kilowatts per hour, which means an overnight charge should be plenty to top off the battery. The charger itself can cost from $500 to $2,500. They require a 240-volt outlet, which newer homes often come with. Installing one in an existing home can cost $1,000 in electrical work—or thousands more, depending on how much futzing is required.

EV owners gush over the convenience of powering up at home and bypassing gas stations. In the United States, an estimated 80 percent of electric-car charging happens at home.[11] That's also the key to a big cost-of-ownership advantage compared with an internal-combustion-engine vehicle. The savings vary depending on many factors, including model type, utility rates, and gas prices in your area. But it's not uncommon for an EV owner to spend just one-quarter the amount on home charging than they would to fill up a comparably sized car at the gas station.[12]

There are also tens of thousands of public Level 2 chargers in the United States, located everywhere from hotels and office parks to grocery stores and libraries. Rates on these public chargers typically run owners 30–50 percent more than they would pay at home.[13]

The reality is that home charging solves most needs for most people in most situations. Nearly all the EVs on sale in the United States have driving ranges of at least 225 miles, and 300 to 400 miles has become more common. It's plenty of juice for almost any round-trip commute—the average one in the US is about 28 miles one way—or to last a day of running errands and shuttling kids for even the busiest parent.[14] Most EV owners will get through a typical week without any need to charge away from home.

But the biggest problem with charging—which acts as a natural cap on EV adoption—is a lack of Level 3 fast chargers, the kind that David Jankowsky wanted to install for Mayor AJ. Once an EV owner needs to leave town for a few hundred miles—a business trip a state over or a weekend getaway to the cottage—things can get complicated quickly. When talking heads and policy makers cite the critical need to build out the nation's EV charging infrastructure, they mostly mean Level 3 fast chargers for these longer drives.

The availability of fast chargers along highways is a crucial factor in reducing the dread that creeps into EV drivers as they watch those battery-capacity levels drain while cruising the interstate. Highway driving in an EV is the least efficient mode, and the mileage can dwindle at a sweat-inducing rate. Until road-trip charging becomes almost as quick and straightforward as a typical trip in a gas-powered car—with ample off-ramps to fuel up—a significant percentage of US car buyers will shy away from going electric.

In the meantime, operators of interstate travel centers are ready to serve those EV owners who have extra time to pick out their beef jerky and energy drinks as they wait for the electrons to flow. Companies like Love's Travel Stops and the Buc-ees chain of rest stops are adding electric chargers alongside their gas pumps.

In 2023, China's fast chargers outnumbered those in the United States by about twenty to one.[15] Even when US EV owners find one, things often don't go well. Reliability is poor, from glitchy screens and bad connections to broken credit-card readers. A J.D. Power survey in 2023 found that one in five EV drivers who stopped at a charging station left without powering up at all.[16] And EV owners' satisfaction with the system was declining despite efforts by industry and the White House to accelerate its build-out.

The patchiness and problems of America's EV charging network isn't surprising, given that the network is basically in its toddler stage. EVs have been around for less than a decade in any real volume. It took a few decades in the early 1900s for a network of gas refilling stations to evolve so that drivers rarely ever ran the risk of being unable to find a station when in need.

That's probably why, for a long time, auto executives shrugged when asked whether the lack of an EV charging infrastructure might keep a lid on EV adoption. Yes, they wanted more stations built so future EV buyers would have greater confidence. But the view of the industry was more or less, "Why should we be the ones who have to do it?" It's not like Ford and GM built out the nation's 145,000 retail gas stations. With all the things car companies had to spend their money on—new EV and battery factories, driverless technology, software—chargers was low on the priority list.

When Mary Barra was asked about charging stations after showing off the new Chevy Bolt in 2106, she said flatly, "We're not actively working on providing infrastructure."[17]

One car executive was taking a different view: Elon Musk. Tesla in the mid-2010s was building out a nationwide footprint of fast chargers, branded Superchargers. In 2012, Tesla said that it had opened a network of six Supercharger stations "constructed in secret," including a handful between Los Angeles and San Francisco. Owners could replenish their Model S in about a half hour. Tesla initially partnered with charging companies, but as with most things, Musk decided the company would bring the technology in house and develop its own. Tesla, the company said, "is able to provide Model S owners free long-distance travel indefinitely."[18]

Compared to the spotty, scattershot collection of fast chargers from independent companies, the Supercharger network was a relative panacea for Tesla owners. Often referred to as a "walled garden" approach, the Supercharger stations took much of the uncertainty and friction out of play for Tesla drivers—and only Tesla drivers, as the chargers weren't compatible with other automakers' EVs. Owners knew that stations were optimally located, the chargers would usually work, and the stop would be relatively quick. Of course, even as the network rapidly expanded, the sharp growth of Tesla's sales meant that sometimes lines would form with drivers waiting to top up. Eventually, it was no longer free. Still, Tesla owners in surveys consistently cite the Supercharger network as among their top reasons for purchase.

How a startup was able to build its own network, when even deeper-pocketed car companies shied away, is one of the more remarkable aspects of Tesla's story. By 2023, Tesla had more than 2,000 Supercharger stations and around 25,000 individual charging units, about 60 percent of the nation's total.[19] The company's homegrown charging hardware also used a different style of plug than those used by the other charging companies, one that was universally deemed easier to use and more reliable.

Outside those garden walls, it's a jungle. EV owners are confronted with a mishmash of chargers with different names, varying charger plugs, and a jumble of apps to manage payments. Some EV owners complain of having more than a dozen charging apps on their phone. Sometimes property owners let their service contracts with charger operators lapse, leaving them bricked or in disrepair.

It's a big problem for legacy carmakers like Volkswagen and GM, which are trying to make up ground on Tesla with what amounts to an inferior experience. And it will only get harder as they try to expand EV ownership to the masses. Early adopters tend to be willing to put up with hassles and spend the time solving problems—researching the optimal route for a road trip, for example. Mainstream buyers won't be as patient. Plus, many of the earliest EV owners are affluent enough to have a second or third car in the family available for longer trips. Mainstream consumers need to be convinced that their electric car is a viable road tripper, even if they might only leave town a once or twice a year.

Automakers were staring down a chicken-or-egg problem that threatened to hobble their EV revolution. In 2024, for example, GM was looking to make up for its choppy start on EV production by cranking up production at least tenfold. Buyers of those GM EVs would have to compete for charger availability with owners of other brands that are also rushing to sell more electrics. Or worse for GM: customers might simply not take the chance on an electric, unconvinced the infrastructure is ready.

By then, Barra had walked back her view that GM didn't need to be part of the solution. Once again following a path Musk had laid, GM had begun spending $750 million on charging infrastructure. Still, it wouldn't be a proprietary setup like the Supercharger network. It was unlikely to spur the confidence that Tesla's network did. Would a future car shopper considering a Chevy Blazer EV or GMC Hummer EV know they could count on finding a charger? GM and every other automaker weren't sure.

Executives from the old-guard car companies looked longingly at Tesla's walled garden. Some wondered if there was a way in.

. . .

In 2022, Ford CEO Jim Farley was making the three-hundred-mile drive from Lake Tahoe to Monterey on California's central coast in a Mustang Mach-E with his family. His range was dwindling. He came upon a Tesla Supercharger station—off limits to a Mach-E owner—and kept going. "What the hell, Dad!" one of his kids said. "Why can't we stop there?" Farley answered, "Because that's Tesla."[20]

The experience left Farley exasperated. And he knew Ford's EV customers probably often felt the same. Ford had already approached Tesla about a potential deal that would grant Supercharger access to owners of Mach-Es and F-150 Lightnings. Tesla said no. But Farley figured it might be time to make another run at Musk. And not only because of the humiliation of having to drive past that Tesla station with his incredulous kids giving him a hard time.

Musk had talked for years about the possibility of opening up Supercharger access beyond Tesla owners. It had already done so in some European markets. But now there was a financial incentive for him to do so in the United States. Tucked into the $1 trillion infrastructure bill that Congress passed in late 2021, the Biden administration was readying plans to dole out $7.5 billion for EV chargers, most of it to go toward highway fast chargers. The White House made it clear that Tesla could qualify for some of that money, but only if it agreed to allow some visitors into its garden.

. . .

"Hey, Elon? Are you there?" Jim Farley's voice called out.

After a long pause. "I am. Um . . . " Musk answered, trailing off. It was May 2023, and the CEOs had joined for an audio chat on Twitter Spaces. After some awkward back-and-forth and uneasy laughter, Musk asked Farley: "Um, so, would you like to make the announcement?"

Farley broke the news that, starting in 2024, all of Ford's EV customers would get access to twelve thousand Tesla Superchargers, or about half of the overall network. Maybe even more significantly, Ford said that it would be adopting Tesla's connector design on future models, starting a few years out. It was a strong signal that the entire auto industry, which had bet on a different charger plug, called CCS, would now be migrating over to Musk's version. In the interim, Ford customers could buy adapters to power up at Tesla's stations once the agreement took effect.

"We don't want the Tesla Supercharger network to be like a walled garden," Musk said. "We want it to be supportive of electrification and sustainable transport."

Farley showered praise on Tesla. "We love the locations, the reliability, your routing software, the easy use of the connector," Farley said. "It's pretty amazing what you and your team has done for the customers."[21]

For Ford, the deal would more than double its customers' access to fast chargers overnight. It also let Ford bask a bit in the Tesla glow. For Musk, there was some downside risk. Some Tesla owners immediately took to social media to vent that their exclusive stations were about to be inundated with brutes in Ford Lightnings. But there was financial upside for Tesla. It stood to make some incremental revenue from an influx of Ford customers paying to pull juice from its chargers. More importantly, Tesla's move appeased the White House and positioned the company to qualify for some portion of the billions being made available for fast chargers.

It soon became clear that Musk's promiscuity wasn't confined to Ford, despite his chummy exchange with Farley. The deal jolted execs across town at GM, who already had been talking with Tesla about Supercharger access. Two weeks later, GM trotted out an agreement nearly identical to Ford's. Owners of the Cadillac Lyriq, GMC Hummer, and other GM electrics would also be able to use twelve thousand Tesla chargers, and GM would switch future EVs to use Tesla-style plugs. Barra even did her own Twitter Spaces conversation with Musk—just as awkward, but much shorter.

Both Ford and GM touted their Tesla deals as consumer-friendly wins. Investors even bid up the automakers' stubbornly depressed share prices. It was, however, also a humbling scene that might have been hard to imagine in another era: GM and Ford, celebrating the right to send their customers to refuel their cars at a machine with a rival's name emblazoned on it. But here they were.

The industry's capitulation would be complete about two months later, when seven automakers announced they would join forces to fund a network of fast-charging stations that would rival Tesla's in size, targeting about thirty thousand chargers. The companies, which included BMW, GM, Honda, and Hyundai, agreed to jointly kick in $1 billion to start.[22] The stations would include both Tesla plugs as well as the CCS-style ones. Even as the industry moves to the Tesla connector, EV drivers in the United States will be living with the dueling hardware for many years.

It had become clear that the pot of money being offered by the Biden administration was enticing more players to take a crack at solving the largest obstacle standing in the way of an EV tipping point. Traditional carmakers, nonprofits, utilities, travel-plaza operators, and entrepreneurs like Jankowsky all were mobilizing for funding. The federal government awarded grants to the states, which decided where the charging stations were needed and ultimately who would get money to install them.

For Jankowsky, the funds flowing out of Washington, DC, made those Oklahoma tax credits look like couch change. "Before we were going into states and begging for 5 million bucks in tax credits," he said. "That has been dwarfed." Jankowsky expects to eventually be awarded around $100 million in contracts from states through the federal program.

The federal push to get behind EV charging is making Jankowsky's leap-of-faith business model suddenly much more viable. The way he sees it, the funding is also giving places like the Plains and other areas of fly-over country a fighting shot at making EVs work in the long run.

"Without that federal funding, there is no EV charging network in the heartland," Jankowsky said. "It can really make the economics work for those rural areas where EV utilization is going to be quite low for quite a long time."[23]

Of course, all of these freshly installed chargers won't matter if the US electric grid can't handle the extra load from a proliferation of EVs. The limit of the grid is one of the main arguments put forth by those with bearish views of the EV story. In short, the US grid can handle a big influx of EVs. But most developing countries, not so much.

A 2019 study jointly funded by the federal government, researchers, and utility operators—a group called U.S. Drive—concluded that "EVs at Scale will not pose significantly greater challenges than past evolutions of the U.S. electric power system."[24] A 2023 analysis by *Consumer Reports* found that the EPA's more-stringent emissions standards through the early 2030s would boost electricity demand only by a manageable 6 percent.[25]

A 2022 study from consultancy KPMG estimated that the United States has the electricity capacity to charge 80 million EVs during overnight hours, when most owners plug in. That same year, there were only about 2.5 million registered EVs in the United States. Still, to make it work, KPMG said that some investments would be needed to ease bottlenecks, and utilities would need to efficiently manage loads. Globally, though, the

challenge of generating and delivering the power needed for a broad transformation to EVs seems monumental: more than half the population—
nearly 4 billion people—"live in developing economies with inadequate
electric grids. When that will change, no one can tell."[26]

. . .

As David Jankowsky traveled the back roads of middle America trying
to find takers for his fast chargers, Carter Li was trying to solve another
huge—and in some ways opposite—barrier to EV adoption: apartment
buildings.

Li was a thirty-something consultant for the big firm Deloitte in Toronto. A Montreal native and son of Chinese immigrants, Li always was
something of an environmentalist. He remembers watching *Captain
Planet and the Planeteers* as a kid. He got his undergraduate degree in
biology and his PhD in bioengineering. In college, he did an internship
with the Smithsonian Institution in Panama, where he studied the effects
on snail populations from chemicals used in the Panama Canal to prevent barnacles clinging onto ships. It was only natural that Li would want
to buy an electric car, even back in 2015, when nearly no one in North
America did.

Li was eyeing a Nissan Leaf. First, though, he wanted to make sure he
would be able to charge the car at his condo in Toronto. He figured it
might be a tough sell to persuade his property management company to
take up the issue. Like many apartment buildings, his was decades old.
Most buildings aren't set up to route lots of power into the basement parking area. The power company would need to do a significant overhaul,
including a new electrical panel, which would cost around $10,000.

Li also knew it was unlikely that his fellow condo owners would be
eager to cover these expenses. So Li offered to foot the bill himself. During
many meetings with the property manager and condo association over a
year, he explained his reasoning, how electric-car sales would grow gradually over time, and having the building prepped to handle chargers would
enhance the property value.

Nobody was buying it. The property manager ultimately just didn't want to deal with it. "They were like, 'You know what, this is just too much hassle for us. We don't really think electric vehicles will be a thing. We'll be stuck with all this conduit and infrastructure to maintain. Sorry, we just can't do this.'"[27]

Li was frustrated. He moved to a different building that already had parking-garage chargers and bought a Chevy Bolt. But he kept turning over the experience in his head. Much of the world's population lives in apartment buildings or other multiunit buildings. If his building owners wouldn't even agree to do it for free, how was this ever going to work?

"How is the electrification of transportation going to really be completed if like two-thirds of the world can't practically own an electric vehicle?" he thought to himself. "I just felt like this was a problem worth solving."

Li hadn't ever considered starting his own company. Early on he was interested in a bioengineering job that would help the environment, maybe something in the field of carbon sequestration. But he later concluded those jobs would take decades to really make an impact. Management consulting was a shortcut to work that might translate into meaningful change, he figured.

If Li had one takeaway from his three years as a consultant, it was that no matter what it seemed like from the outside, nobody in the business world had things figured out. Working across industries—energy, financial services, pharmaceuticals—he encountered a lot of people who he sees as just sort of winging it. There were decisions based on bad or incomplete data, managers forging ahead when they shouldn't—or standing pat when they needed to act. It was a good reference point from which to take this frustration with his condo building and develop a business plan around it. If everyone was winging it, why couldn't he?

Li started out with some market research. He visited some Tesla owners' group meetings. At that point, Tesla was the only EV brand big enough to spawn enthusiast groups. The results weren't encouraging.

He learned that early adopters were mostly affluent people who lived in single-family homes. "Everyone was like, wait, you own an EV in an apartment building? That's crazy." He didn't get much further with property managers and real-estate developers. "They more or less laughed me out of meetings," Li recalled.

He was determined. His vision was a company that could serve as a plug-and-play solution for building owners to manage the infrastructure upgrades with the utility, bundle residents' projects, and divvy up billing. He quit his job and began scrounging for grants and stipends available for entrepreneurs. The firm's first big break came in 2018. Li had applied for a grant from the Ontario government that awarded startups focused on reducing carbon emissions. He was working in a shared office space with a handful of employees when the email came in that said his firm had been awarded a $250,000 R&D grant. "I just remember jumping up and down screaming," he recalled.

Li was able to load up on software developers to optimize the charging of various apartment dwellers. Some owners might need only to top up from a 70 percent charge. Others might be scraping empty. Li's company, which he named SWTCH, developed software to manage those loads and make sure everyone ends up with their full charge when they need it. Then, there are times of the day or night that are the most economical to charge, when the utility charges the cheapest rate. The software manages all that too. The building housing the charger pays a flat fee for Li's service.

The next year, 2019, Tesla's Model 3 had taken off and EV sales were rising. Li began hiring more software developers and salespeople. The next year, 2020, SWTCH raised about $1 million, mostly from clean-tech investors. In 2022, it raised $13 million more from venture capital firms. In April 2024, it pulled in $27 million.

SWTCH was on pace to finish 2024 with about 160 employees, nearly 20,000 chargers installed and $40 million in annual revenue.

Every so often, Li will get an email or a call from a property manager, one of the ones who had dismissed him or laughed him out of the room a few years ago. He's more than a little satisfied by those calls. "They'll say, 'Hey, I've had like four tenants ask in the last few months whether we'll be installing chargers. Do you mind coming out?'"

6

Lightning Strikes

n early 2020, around when Mary Barra and her GM team were at the dome showing off a future fleet of silk-sheet-covered EV prototypes, Ford Motor Company's stock price was buried at a multidecade low. It was the dark early days of the pandemic, when North American factories were abruptly shuttered, and car companies instantly began bleeding cash.

But in less than eighteen months, under rookie CEO Jim Farley's leadership, Ford's share price had jolted back to hit a record high for the 118-year-old company. Wall Street—then infatuated with all things EV—had finally started to buy into Ford's plan to take on not only Tesla, but also its traditional rivals, many of which just a few months earlier were perceived to have lapped Ford in the race to an electric future.

And of these, General Motors was the bitterest of rivals. For years, GM had been getting all the love from investors. As early as 2017, Mary Barra had begun making decisive proclamations about being all in on EVs. She fed Wall Street's infatuation with the anticipated massive disruption to the transportation sector, and all of the money-making possibilities tied to it: visions of quiet, clean electric cars—many of them operating autonomously as robotic Ubers in a highly profitable network of US cities. Barra tantalized investors with a new, oft-repeated slogan that seemed to crystallize those utopian visions: GM was working toward a future of zero emissions, zero crashes, and zero congestion.

Sure, Tesla's stock valuation sat gaudily at several times that of Ford's and GM's combined. But Barra had cemented GM's premium over Ford for nearly a decade.

By December 2021, that had flipped. Farley's moves to get Ford into the game on EVs finally got investors' attention. And Ford executives were about to put an exclamation point on it. Across town, Barra had spent weeks preparing for her keynote speech at the annual consumer-electronics show, CES, an early January pageant of the latest, most cutting-edge tech held in Las Vegas. The stakes were high. Barra was to take the stage to reveal one of GM's most important new vehicles in decades: an electric version of the Chevrolet Silverado, the company's top-selling model and its biggest profit engine. For months, the GM team had been grating under the praise being heaped on Ford's electric F-150 Lighting pickup.[1] This was their chance to show that GM's approach—a truck designed from the ground up as an electric vehicle—would blow away Ford's makeshift retrofit of a gas truck.

The morning before Barra's big appearance, Ford PR staffers began quietly reaching out to automotive reporters to line up calls with Darren Palmer, a jovial Brit in charge of Ford's electric-vehicle programs. In a series of interviews, he said Ford was doubling factory output of the Lightning—again, the second time in just four months—in response to overwhelming demand. "The feedback has been immense," Palmer gushed. "Everywhere I go, from congressmen to the guards at the door, everybody is talking to me about the Lightning."[2]

The message was clear and intentional. The Ford team knew Barra would take the stage to wax on about how the Silverado would deliver performance that competitor EVs couldn't match. And she would be right: because of GM's from-scratch approach to EV development, engineers were able to fit more batteries into the Silverado, delivering a longer range, faster charging times, and more interior space because of its efficient layout.

But Ford's media blitz drove home a simple point: Ford will have sold more than 100,000 Lightnings, Palmer implied, before the first Silverado rolls off GM's assembly line. Wall Street investors gobbled up Ford's news. Its stock leaped 12 percent that day, to close at its highest level in about two decades.

A national Covid-19 surge over the holidays prompted Barra's team to cancel her CES appearance. Instead, she filmed her keynote at Detroit's historic Fox Theatre, beaming the video into the Vegas conference. Barra, who traded her trademark black leather jacket for a crisp white blazer, played to the tech crowd, making the case that GM is no longer a car company but a "platform innovator," ready to pump out a string of electric, high-tech vehicles that can connect to owners' everyday lives.

Then a gleaming, bright-blue Silverado rolled onto the stage, with its futuristic-looking face streaked in LED lights, its silhouette cutting a sleek, taut profile, like a pickup truck that had been hitting the gym. The truck, Barra said, would have a driving range of four hundred miles and could be charged up to a hundred miles in just ten minutes—stats that handily beat Ford's Lightning. "No other automaker today matches the depth and the range of GM's growing, all-electric portfolio," Barra said. "Make no mistake, this is a movement."[3]

GM's stock fell 5 percent by the end of the day. Wall Street analysts theorized that Ford's head start was more appealing to investors than GM's admittedly impressive specs. "The optics to many investors," RBC Capital wrote in an investor note, "will be that GM is lagging behind Ford."[4]

In the following weeks, Barra and her team made the rounds with investors and analysts to explain how GM would rapidly scale up EV factory output. But the conversation kept coming back to Ford. Why would the Silverado trail the Lightning by more than a year? Could the launch date be moved up? The undertone was clear: how could GM—the company that had declared itself "all in" on EVs years earlier—have let Ford take the lead in the crucial pickup-truck market?

"All the questions about Ford," one person on her team said of the meetings, "were starting to get under her skin."[5]

. . .

The seeds of Ford's comeback were sewn years earlier, some four thousand miles away. In 2017, Jim Farley had been stationed in Cologne, Germany, running Ford's European business. Ford's EV strategy at the time—or lack thereof—gnawed at Farley. The longtime marketing executive had a front-row seat to the nascent but distinct movement toward electric cars in Europe. The fallout from Volkswagen's emissions-cheating scandal was still percolating through the region's car market. Governments in Europe were demanding that consumers wean themselves from their beloved diesel cars, which then accounted for roughly half of sales.

Farley was one of the few traditional auto executives who was taking Elon Musk's car company seriously. Tesla had been straining to ramp up production of the Model 3, its first mainstream model. Musk was famously sleeping on the floor of Tesla's lone factory, in Fremont, California, to help push through manufacturing bottlenecks, which he dubbed "production hell." But from Farley's vantage point in Europe, Tesla was getting traction. He was seeing more Teslas popping up on European roads, and on the region's sales charts. Around 3,500 Teslas were sold in Norway in 2016, he noticed. A year later: 8,000.[6]

"I was like, 'Oh my god. Tesla is selling all these vehicles in the Netherlands, in Norway. Now the French are starting to sell electric cars,'" Farley would say later. "This was my marketplace, and all the sudden it's like, 'Holy cow, this is moving in Europe.'"[7]

Auto-industry executive suites are filled with financial types, manufacturing experts, even engineers, like Barra. In general, though, they're businesspeople. They all drive their company's vehicles and have their favorites; some might even be comfortable tinkering under the hood. Most, though, are not true car enthusiasts. Farley's passion for cars, and the car business, is different. He races cars, collects cars, wrenches on vintage cars, and builds miniature car models. He can engage in stories about cars and the industry for hours. And the frank emotion with which he spoke about the business—to employees, Wall Street analysts, dealers, journalists—was unusual, and often left them wanting more. In a 2008 interview with the *New York Times*, Farley used an analogy to describe Ford's battle to avoid bankruptcy at the time: "I cut this finger playing with a model plane when I was 12 or 13," he told the paper. "I didn't really feel it. It was like, that's weird. Then the whole finger came down and I could see the white of my bone." He continued. "Some cuts leave a little scar, and some cuts go to the bone. Ford's experience in the last 10 years went to the bone. My hope is that everyone at Ford never forgets what we went through."[8]

A first cousin of the late comedian Chris Farley of *Saturday Night Live* and *Tommy Boy* fame, Jim Farley was born in Buenos Aires in 1962. His father was a banker, and the family relocated many times. As a ten-year-old boy in Connecticut, he had a paper route that just happened to include a Ferrari dealership. The young Farley would ride his Schwinn Stingray bicycle to the dealership and wander down to the basement, where Italian mechanics worked on their old race cars. He would chat them up about the Ferrari 512 sports car or famed racer Mario Andretti. "My dad really didn't like cars. He didn't allow me to watch any car racing," Farley said in 2020. "That Ferrari distributor was a really good introduction" to the passions of the car business.[9]

While his upbringing was cosmopolitan, Farley also has Detroit roots. His grandfather, Emmet E. Tracy, joined Ford in 1913 as the automaker's 389th employee. Every Christmas growing up, Farley would visit his grandfather in the adjacent Detroit suburb of Gross Pointe. There, a neat stack of that year's issues of the *Automotive News* trade publication—fifty-two of them—awaited Farley. While the weekly publication was held in high esteem in industry circles as the car business's paper of record, most teenage gearheads surely would have gravitated to sexier car magazines like *MotorTrend* or *Car & Driver*. Farley wanted to devour stories about obscure parts suppliers, dealer strategies, and car-company marketing campaigns.[10]

Farley likes telling the story of when he was a fourteen-year-old while working the summer in California and paid $500 for a broken-down 1966 Ford Mustang. He spent weeks rebuilding the engine by hand, later driving it back to Michigan—underage. He still keeps a photo of himself dressed in all white, down to his dress shoes, leaning against the black muscle car.

But what Farley considers his real indoctrination to the car world came when he was working his way through graduate school in the late 1980s at UCLA, after getting his bachelor's degree in economics from Georgetown. If he was going to work forty hours a week to pay for his MBA, Farley figured, he'd at least aim for a job that would scratch his car-obsession itch.

He knocked on the door of a famed car-restoration shop in Santa Monica owned by his hero: race car driver Phil Hill.

Hill was the first American to win the Formula 1 World Championship, in 1961, and was a three-time winner of the famed 24 Hours of Le Mans endurance race.[11] But he was more than a racer. Like Farley, Hill was fascinated by the industry's rich history and culture. In 1955, Hill won a famous road race at Pebble Beach—*and* won the best-in-show award at the prestigious vintage-car show connected to the race, with his1931 Pierce-Arrow.[12] Hill spoke Italian and was an opera aficionado. His restoration shop, Hill and Vaughn, was considered one of the most renowned restorers of vintage automobiles in the country.

Farley was thrilled when Hill hired him—as a janitor. Once inside, though, Farley set about giving himself an education. On his lunch break, he would chat up the mechanics about car design and engine specs, much like back at the Ferrari dealership years earlier. Being around so many exotic models allowed Farley to immerse himself in the intricate nuance of vehicle design, that combination of proportion, styling cues, and silhouette that dictate, almost subconsciously, why some car designs work and others fall flat. Farley said he learned why, say, this Delahaye is better than that Delage (both French super cars); why an 8-liter Bentley was better than a 6-liter Bentley. "I got my PhD in the car business," he said. "It's where I learned why a beautiful car is a beautiful car."

Eventually, Farley worked his way onto the restoration team. He taught himself to stitch leather interiors so he could do custom jobs for the shop's clients. "All the cars from the thirties were custom made, and people would pick not just exotic colors but materials," Farley said. "Ostrich hide. Frog. You'd have to sew, like, five hundred frog hides together to make a seat cover."

Around that time, Farley was straddling the worlds of cars and business. One summer, he interned on a J.P. Morgan mergers-and-acquisitions team and was exposed to the gaudy sums of money available in the world of corporate deal making. But then he happened to attend a lecture at UCLA by Jim Collins, the author of the best-selling business book *Good to Great, Why Some Companies Make the Leap . . . and Others Don't.* The message encouraged Farley to marry his passion for cars with a career in business. "He said to take a risk," Farley recalled later. "My risk was to join the car business."

After graduating in 1989, Farley was offered a job by Ford to be the financial planner for the rear differential on the F-150 truck. Now, a rear differential is an important component, especially on a pickup truck: it connects to the driveshaft and directs the right amount of power to each wheel. But, Farley thought, a financial planner . . . for one car part? Around the same time, Farley interviewed with a new luxury brand: Lexus, owned by Toyota. He was offered a position as a product planner, which meant he would be part of a team that works on models from conception through production. "In the interview I said, 'You mean, I get to work on the *whole car?*'" With that, he found himself somewhat conflicted, passing on the company that was in his blood for a fledgling Japanese luxury brand.

Farley's marketing career took off at Toyota. His team worked on the soon-to-be stalwart seller RX-300, one of the industry's first so-called crossovers, which took on the body shape of an SUV but used the frame of a car, unlike the truck-based SUVs of the day. The setup allowed for a more elegant design, with softer edges and a smoother ride. Farley applied his learnings from Phil Hill's shop about the importance of getting a car's profile right. "When you come up with a new silhouette like that, you can really catch the competition sleeping," he would say later. "It took them five years to catch up. I learned a big lesson there." Farley's team emphasized women as the target buyer, a rarity at the time. It stayed on that theme when it created a retail strategy for Lexus with a focus on a low-pressure, less-intimidating environment at dealerships.

Farley rose through the marketing ranks at Toyota. He helped launch a new, youth-oriented brand, Scion, and built it to 100,000 vehicle sales within two years. For Scion, Farley made it a point to hang out with trend-setting twenty-somethings, studying their consumption habits and brand affinities. "You need to love your customer, feel their joy, understand their pain," Farley told author Bill Vlasic for the 2011 book *Once Upon a Car: The Fall and Resurrection of America's Big Three Automakers—GM, Ford, and Chrysler.* "You have to get so close to them you can smell their breath."

Farley eventually took over the Lexus brand in North America. By 2005, he landed on Ford's radar back in Dearborn. Bill Ford arranged to have a clandestine introductory meeting that January with Farley at the Ford family's Detroit Lions headquarters. Farley was impressed with Ford—who was then the company's CEO—but wasn't looking to leave Toyota, which had been kicking the Blue Oval's ass in market share for years. It didn't help that the meeting was in the frigid gray of January in Detroit. "I'm from Santa Monica. I work in Torrance [California]. I go to Toyota City in Japan," Farley thought to himself, according to Vlasic's book. "That's my auto industry. This is just so . . . old."[13]

Two years after Bill Ford started courting Farley, it was Alan Mulally who closed the deal. Mulally, a Kansas-bred aerospace engineer and one-time head of commercial airplanes at aerospace giant Boeing had made a surprise jump to lead Ford in 2006, allowing Bill Ford to step away from day-to-day duties. During an eight-year tenure, the ebullient Mulally built an almost legendary legacy for improving Ford's operations and globalizing the company.

Farley had spent his entire career at a company whose mission was in large part to capitalize on all the things the Detroit car companies were getting wrong, gobbling up US market share with a keen focus on quality, fuel efficiency, and manufacturing expertise. Mulally spoke that language. The two discussed *kaizen*, a core principle of Toyota that translates to continuous improvement, one of the underpinnings of the automaker's legendary manufacturing system. They talked about plan-do-check-act (PDCA), another mainstay Toyota philosophy for problem-solving.

"All those things I grew up with as basic tenets at Toyota, Alan believed in too," Farley said later. "When I met him, it was pretty instant. I said, 'I want to join this company.'"

Farley joined Ford in 2007 and soon became a core part of Mulally's team during a resurgent time at Ford. When the other Detroit automakers were suffering the ignominy of bankruptcy in 2009, Ford was growing and leveraging its status as the only US automaker that didn't undergo a federally backed restructuring—not a bad claim to tout when you're trying to sell pickup trucks in the heartland.

Farley turned his intensely competitive streak back to focus on beating the company that he'd loved for nearly two decades. Toyota—even more than the despised General Motors—was the number one target. And one figure became the driving motivation for Farley in that effort: $1,500.

It came from some marketing focus groups, in which prospective car buyers were asked to name the price they would pay for cars—some with no brand visible and some with the badges displayed. On average, buyers were willing to pay $1,500 more for the vehicle with the Toyota badge on it. Fifteen hundred dollars for a brand name! It focused Farley on the task; he kept a placard in his office with the figure printed on it: $1,500.

Farley again got to apply his learnings from Phil Hill's restoration shop in California when helping Ford's product-development team lay out a new look for Ford. Its vehicles took on more stylish, European profiles and driving dynamics. Meanwhile, Mulally's "One Ford" strategy standardized the mechanical layouts of Ford's vehicles across global markets, which eventually grew its economies of scale, improving manufacturing quality and the bottom line. Ford even began chipping away at Toyota's sterling reputation for green cars, in part through its own line of hybrid vehicles, helping Ford shake off its image as an old-school Detroit purveyor of gas-guzzlers.

Ford moved up the rankings for vehicle quality and became seen as a more-premium player among mainstream car brands. The brand commanded higher prices for its cars. Farley said that the $1,500 gap evaporated.[14]

· · ·

It's a decade later now, 2017, and Farley is in Cologne, stewing over Tesla's fast-rising profile in Europe. Jim Hackett was recently named Ford's CEO, replacing Mark Fields. Fields, a New Jersey native and Ford lifer, had replaced Mulally after waiting in the wings for years, but his run was anticlimactic. He struggled to revive Ford's flagging stock price or chart a growth strategy. The same day Hackett replaced Fields, in May 2017, Farley was summoned back from Germany to Dearborn.

Farley's new title, president of global markets, put him in charge of, or in contact with, just about every facet of Ford's business. Leaders of North America, Europe, and Asia reported to him. Farley took over the familiar role as head of marketing and would oversee the Lincoln brand. He was also tasked with charting Ford's course on electrified and autonomous vehicles.

Soon after his return to Michigan, Farley caught wind of a new, fully electric prototype in the works inside Ford's engineering center in Dearborn, news that hit him like a bucket of cold water to the face: there had been a fully electric car under development, and nobody thought to loop in the head of Europe, where an EV revolution was underway? "I saw the first evidence that this was moving in Europe, but they didn't care to ask me?" Farley asked himself, incredulous. Furious, he demanded to see the car.

A few dozen designers, engineers, and marketing types gathered for a hastily called design review at Ford's studios in Dearborn. Farley listened silently as the development team walked him around the vehicle, a midsize SUV about the size of a Ford Edge. Farley stood silently for what seemed like minutes. He thought it looked pedestrian, nondescript, like any utilitarian gas-engine SUV on the market at the time.

"It looks like a Prius. That's a joke. What are we doing?" Farley recalls telling the stunned development team. "Nothing about this is compelling. Please stop right now."

The program's engineers gently pushed back. The project was far beyond the so-called design freeze, the point at which changes to the exterior styling and vehicle's dimensions—what designers call the hard points—were set in stone. Major changes at this stage are akin to reopening the patient after surgery. It would set back the model's timing by at least six months, they said.

"Fine, take a six-month delay. This is not going to fly," Farley recalls saying. "I will never approve this.'"[15]

Such a harsh rebuke of a prototype was unusual. But it belied Farley's evolving viewpoint that traditional automakers, including Ford, had botched their EV strategies for a decade. They treated electrics simply as a regulatory box that needed checking. They viewed the potential market mainly as earthy granola eaters who wanted an electric for environmental reasons, or penny-pinchers just looking avoid spending money on gas.

Thus, most of the EVs that went on sale over the previous decade had been small hatchbacks, outfitted with just enough battery range to get the driver around town for a day—think a Nissan Leaf or Fiat 500-e. Adding more range meant bigger battery costs, which already were so expensive they made EVs money losers.

Ford fell into that trap along with many other traditional carmakers. It marketed an all-electric Ford Focus small car from model years 2012 to 2018, with its paltry seventy-six miles of driving range—less if the owner had the nerve to turn on the heater in winter, which sucked down the battery even faster.[16] Ford had sold about 8,700 of them over about seven years. The company sells about 8,700 F-150 pickup trucks in less three days.

Tesla turned that vanilla thinking on its head with its stylish, lightning-quick cars that earned a fervent, growing fan base. Executives at places like GM and VW were noticing, but remained skeptical that Tesla could make the transition to mass production.

Farley, on the other hand, was fixated on the template that Musk had set. Why not make an EV that would wow people, regardless of whatever power source happened to propel the wheels? Ford already had a built-in advantage: generations of loyalists for nameplates like Mustang and F-150. Ford, Farley thought, needed to exploit that passion.

"From an electrification standpoint, we are leaning into our strengths," Farley told a reporter at the Detroit Auto Show in early 2018, just a few months after that tense meeting on the boring Ford Edge-like electric SUV. He even dropped a hint of what might be to come. "We can fit a battery pack between the rails underneath our internal-combustion F-150," he said with a smile, hinting at his new strategic direction while stopping short of committing the auto-industry cardinal sin of divulging future vehicle plans to the press.[17]

Soon after he hijacked the electric SUV project, Farley began pressing for the EV to take on design elements inspired by the sporty Mustang, which boasts Ford's most passionate fan base. So designers lengthened the wheelbase to give the vehicle a more athletic stance. The grille was revised to an aggressive, flat-oval shape. Other unmistakable Mustang styling elements were added: taut, muscular haunches, and the iconic three vertical bars on the taillights.

It wasn't unusual for a car brand to transfer styling cues like the taillight design from popular models to the rest of the lineup. Executives at GM's Chevy brand have long said that there's a bit of Corvette in every Chevy model. But Farley wanted to take things a step further: He soon began agitating to affix the Mustang name to the new model: the Mustang Mach-E.

To some inside Ford, the plan was radical and blasphemous. The roar of a pulsing gas engine is in the Mustang's DNA. Plus, to put the hallowed nameplate on a sport-utility vehicle—a body style so utilitarian that the word was part of its name—would be a slap in the face to Ford's most loyal and passionate customers, some Ford insiders warned. Farley didn't think so, and he pressed the issue. And unlike a lot of auto executives, he had the car-guy credibility to not get laughed out of the room.

Ultimately, Farley knew he had a good case and knew he would have to get buy-in from just one person: Bill Ford. The great-grandson of company founder Henry Ford—and now the company's executive chairman—has owned dozens of Mustangs. If Farley wanted to know whether transferring the Mustang name to an electric SUV would piss off the super fans, he needed to start with this one.

Farley's team approached Bill Ford with the idea—carefully. They didn't spring the name on him right away. Instead, they used the term "Mustang-inspired" to describe the SUV's styling, and they said it would carry the Mustang DNA. A few minutes into the meeting, Ford got the hint.

"You guys aren't telling me you want to call this a Mustang?" Ford asked. After a few evasive answers, Ford cut them off. "No, I'm sorry," he recalled saying later. "I don't want to hurt the brand. This is not going to be a Mustang."[18]

Eventually, Farley's team presented its case during a high-stakes presentation inside Bill Ford's wood-paneled office at company headquarters. They came prepared to make the case that the SUV's performance traits were worthy of the pony Mustang emblem. The base version of the car would have more than 330 horsepower, good enough to beat a Porsche Macan, the executive chairman was told. The higher-performance GT version? The thing will get 460 horses and outrun a Porsche 911.

The specs were intriguing enough to move Bill Ford off his "definite no" to a solid "maybe." But before he went any further, he had to drive it. Maybe Farley was hoping for that, knowing the oomph the car would give him. Bill Ford took a quick spin in a prototype at the company's proving grounds and emerged from the car flashing a double thumbs-up. The Mustang pony could go on the electric SUV.

Farley felt vindicated that his commandeering of the EV SUV project paid off. But that experience—how close the team had come to botching the design in what was to Farley the most predictable way they could botch the design—made him uneasy. He was left with a lingering sense that, going forward, Ford needed another way to develop its next EV entries. He was unconvinced that the current Ford development machine—thousands of designers and planners and engineers that had honed the process with gas and diesel vehicles for decades—was equipped to pivot.

"We cannot ask the normal Ford," Farley would later say, "to develop this first generation of electric vehicles."[19]

He appointed a skunk-works crew of engineers and designers to think big about EV concepts that would turn heads. The group, dubbed Team Edison, was assembled with iconoclasts and mavericks, people Farley figured could power through Ford's corporate slog and do things differently. It set up shop in a newly renovated, three-story, historic Beaux Arts building in Corktown, Detroit's oldest neighborhood, a few blocks away from the remnants of the old Detroit Tigers baseball stadium. Bill Ford's ancestors were among the Irish immigrants to settle in the area, which later also drew an influx of Mexican and Maltese immigrants. Hemmed in by downtown, the Detroit River, the Ambassador Bridge to Canada, and ribbons of highways, Corktown has a thriving restaurant scene, set against a scruffy, industrial backdrop. Before showing the world the electric Mustang, Farley was pressing this team to think about the next EV project.

The idea of an electric pickup had been kicked around already. But for Farley, the concept really took hold during a 2017 conversation with an unlikely source: a thirty-something portfolio manager in Southern California named Bobby Stevenson. He was a automotive analyst and managed some funds for financial giant Franklin Templeton. The energetic Californian had spent plenty of time with automotive clients in Detroit over the years. He had known many Ford executives, but he especially hit it off with Farley. They were both UCLA alums. They both spoke the same car-culture language. Stevenson, though, was mostly a truck guy, who owned multiple pickups and, in his formative college years, would tool around the west side of Los Angeles in his cousin's truck—an early-2000s 2004 black Ford Lightning, a high-performance pickup truck with a massive, supercharged V-8 engine. It was no off-roader: the low-to-the-ground rocket enjoyed a cult following for a few years among urban cowboys looking to make noise and draw looks around town on weekends. Ford phased out the Lightning a few years later, a short-lived fate not uncommon for specialty models.

The Lightning had been on Stevenson's mind when he reached out to Farley in 2017. Stevenson wanted to see car companies go electric much more aggressively and thought an electric truck would be a good idea. He also had an ulterior motive: his firm was invested in Proterra Inc., an electric-bus startup that wanted to build lightweight municipal buses and anchor them with massive battery packs placed under the floor.

Stevenson figured he might be able to persuade Farley to consider a limited-edition electric pickup truck—make a few thousand of them, priced over $100,000—and maybe Proterra could be involved by supplying battery packs or other components. Stevenson knew Farley would be worried about whether a battery could really provide enough range for a large pickup, so he had his pitch ready. "I'm 100 percent certain that if you can put enough kilowatt-hours of batteries under a 30,000-pound municipal bus," Stevenson told Farley, "you can build one that would give you the range and ruggedness you'd need for the F-150."[20]

Stevenson figured the Ford team would be scared off by the idea of slapping batteries into a truck that some owners would take splashing through mud and over rocks. He was ready for that, too. He explained how well fortified the Proterra buses were manufactured. Officials in New York City, he explained to Farley, required Proterra to drop manhole covers onto their battery packs and shoot bullets into them to be convinced that they weren't a safety hazard during, say, a hostage situation on municipal bus.[21]

"I promise you a battery pack can be built that is commercial grade," Stevenson told Farley.

Plus, Stevenson played up the serendipity of the Lightning name being resurrected for an EV. "The F-150 Lightning? For an electric truck? Are you kidding me?" Stevenson told Farley. "This is such a no-brainer. You own this trademark already."

Ford had a window, Stevenson noted, before Tesla would be coming out with an electric truck. "Just source some battery packs, put them under an F-150, and get them out as fast as humanly possible. This is doable."[22]

Farley was intrigued enough to grab some Team Edison members and head off to Proterra's Bay Area facility to take a look. Some Proterra work was done in a nondescript building off the 101 Freeway, in the shadow of San Francisco's airport. The front housed Proterra's offices, looking like any tech company's setup of desks and workers pecking away on laptops. Out back, though, was a small factory where employees worked on massive battery packs that would eventually go into the floors of giant custom-built bus chassis, made from lightweight composite and carbon-fiber materials. With the weighty battery pack, the buses tipped the scales at around thirty thousand pounds, the equivalent of five big pickup trucks.

Farley would later call Ford's visit to Proterra "a huge light-bulb moment." When he got back to Dearborn, he grabbed Sherif Markaby, who was running Ford's EV programs at the time. "Sherif, we're gonna have to do an F-150 Lightning," he told him. Markaby responded: "Yep."[23]

The Ford team got to work right away and didn't stray far from Stevenson's advice. Engineers never considered starting with a from-scratch platform, like GM did. They would never have been able to sink the money into a stand-alone platform, which would take longer and add hundreds of millions of dollars in cost when there was plenty of uncertainty about future demand.

"We sat around and said, 'Who really wants an electric truck?'" the EV engineer, Darrin Palmer, would say later. "We don't know."[24]

Ford effectively took an engineering shortcut with this project. Ford makes nearly 1 million internal-combustion F-150s annually; it's been the best-selling US vehicle for decades running. The most cost-effective way to electrify the truck would be to strip out the engine, transmission, exhaust system, and other guts, and insert a big battery pack and connect it to motors. Use what you have. Only change what you need to.

A few months later, CEO Jim Hackett wandered over to Ford's engineering center, around the corner from corporate headquarters, to see what his team was cooking up. At the time, more of the buzz inside Ford was around the Mustang EV project, which was much closer to maturity, scheduled to be shown the following year, in 2019. Hackett watched bemused as a prototype electric F-150 crawled out of a side door and into the parking lot.

"Do you want to drive this around?" one of the engineers asked the CEO.

The prototype looked like a regular F-150, but with a mess of cords and wires protruding from under the hood. "They basically jury-rigged it so it would work as an electric. It was a mess—like your worst teenage car, minus the beer cans," Hackett recalled.

Hackett was tentative. He climbed in and, for a few minutes, hesitated to take the steering wheel, worried he might unplug something or short it out. "I'm sitting there treating this thing like it's crystal. And they tell me, 'You gotta hit it.' So I took off." The instant thrust of the dual electric motors pinned the six-foot-three-inch former University of Michigan lineman back against his seat. His stomach felt like it had jumped into his throat.

"I turned to the engineer and said, 'I used to own a turbo Porsche Cayenne, and this is better than that,'" Hackett recalled. "And this is the frickin' prototype!" Hackett went up to his office, across the hall from Bill Ford, and gushed to the executive chairman about his experience. The Ford scion, who at the time was fixated on the Mustang Mach-E project, just smiled. The idea of a creating a credible electric version of Ford's biggest money maker was an enticing one for the environmentally conscious executive.[25]

As those projects percolated inside Ford's vast product-development campus in Dearborn, Hackett spent much of the next year fending off questions from the outside about his fuzzy EV strategy, mindful of not revealing future product plans. Hackett and his team would reference a forthcoming "Mustang-inspired SUV," but didn't divulge that it would carry the iconic name and the pony. Lightning was still under wraps.

Across town at GM, Barra's team was making plenty of noise. GM had made big proclamations of having dozens of EV models in a few years, getting into the battery game, steering away from gas-engine development, and funneling capital to EVs. Wall Street analysts were growing louder: GM was killing it. Ford was asleep at the wheel on EV strategy.

Ford finally got to announce to the world that it was wide awake on EVs in November 2019. Farley, Bill Ford, the entire Team Edison, and just about every company executive traveled to Los Angeles for a giant bash to unveil the Mach-E before more than a thousand people inside an airport hangar—a block away from Tesla's design center.

"That's a coincidence," Bill Ford said, flashing a smile.[26] The company scion sat down on stage before the car's reveal with Golden Globe–winning actor Idris Elba, who, before he became famous in HBO's *The Wire*, worked the night shift at a Ford assembly plant in East London, where his dad had also worked for twenty-five years.[27]

"I've driven it, and it's a rocket ship," Ford told Elba, adding that the Mach-E would zip from a standstill to sixty miles per hour in just over three seconds, on par with the fastest-ever gas-powered Mustang. "What I'm most excited about is the lasting impact and what this means for the future of the Ford Motor Company."

Two Mach-Es—one red, one blue—rolled onto the stage amid flashing lights and a booming performance of Macklemore & Ryan Lewis hip-hop anthem "Can't Hold Us" by the Detroit Youth Choir. Afterward, automotive journalists were invited to zip around in prototype Mach-Es on the runways outside the hangar.[28]

A few months later, in March 2020, Farley was elevated to become Hackett's number two and clear heir apparent to the CEO job. He had won a horserace with longtime Ford executive Joe Hinrichs, an Ohio native and manufacturing expert, who had risen through the ranks to president.

Hinrichs was affable and well-liked. He's credited with championing what became one of Ford's biggest product hits in years: the return of the

Bronco SUV in 2021, after a nearly quarter-century absence. But Hinrichs was lukewarm on the growth bets being bandied about the car business at the time—electrics, driverless cars, or the confusing so-called term "transportation mobility cloud" that Hackett sometimes referenced. Ford had bought a shuttle service and even an electric scooter company. It invested $1 billion in a self-driving-car startup. Hinrichs was skeptical of all of it.

Farley, meanwhile, was more attuned to Bill Ford's progressive bent. It was clear to Bill Ford that the auto industry was on the cusp of massive disruption. During the three years since coming back from Europe, Farley had spent much of his time in Silicon Valley. He visited tech companies like Salesforce and Google. To Bill Ford, Farley had shown the intellectual curiosity to lead Ford through that disruption.

Farley took over the top job in October 2020 and immediately began laying out an EV road map. He nearly tripled Ford's planned spending, to $30 billion by mid-decade. Farley made the case that Ford was best positioned for a big move into EVs because it dominated the market for selling trucks and vans to business and government customers, and those users would be most intrigued by the cost benefits of going electric.

In May 2021, during a boisterous party at Ford's historic River Rouge factory near its headquarters, Farley ripped off the sheet covering the F-150 Lightning. The truck could scream to sixty miles per hour in four seconds, he declared, and packed enough electrons to power your house for days during a blackout. The automotive press corps fawned. Two hundred thousand pre-orders poured in, and that was before Ford slapped a shockingly low price tag on it: $39,974 to start, at a time when most EVs were priced north of $60,000.[29]

It seemed clear that Jim Farley had, at the least, gotten Ford in the game. He and his team had turned the tables on GM, generating huge buzz for the Lightning. He visited Jay Leno's garage for a walkaround of the truck. Ford commissioned *Tonight Show* host Jimmy Fallon to perform a song about the Lightning's cavernous front truck, which he dubbed "Junk in My Frunk."

But Farley knew all the pomp was papering over some harsh realities. While Ford might have leaped up on the EV sales charts, the company was still behind GM and others on putting in place the building blocks to ramp up EV production in a serious way. GM and VW already had battery-cell factories under construction and dedicated EV platforms that could serve as the basis for EVs of all shapes and sizes. Ford was a few years behind on that. And Farley needed no reminders that Tesla—for years the source of skepticism for Detroit executives—had become a juggernaut. Elon Musk's company had just opened its second factory, in Shanghai, and was nearing completion of two more, in Berlin and Austin, Texas, extending its lead as the world's dominant EV company.

Farley had a plan to attract talent that would help Ford close the gap on Tesla.

7

The Race for Rivian

Drive about 550 miles northeast of Wagoner, Oklahoma, up I-44 past St. Louis and on to I-55, and you'll land in Normal, an unexpectedly vibrant college town in the farmlands of central Illinois. On the west end of town, a factory sits hemmed in by cornfields. From the road, it looks big, but far away and not out of place, a clean white (gleaming white on a sunny day) structure that rises modestly above the flatness. Get closer, though, and you start to sense just how massive it is—more than half-a-mile long and some eighty acres of space inside.

This is the flagship factory of electric truck startup Rivian. It was here one evening in September 2021 that CEO RJ Scaringe gathered a few dozen of his longest-tenured employees. They met in the "customer engagement center," where buyers arrived to take ownership of their new trucks—at least those willing to make the pilgrimage to the plant. Neatly appointed with tasteful, blond-wood tables, bookshelves, comfy chairs, and many plants, the room was an oasis inside the sprawling, clattering plant humming with the noise of assembly. The factory had stuttered to life just a few week earlier, churning out what would be the first fully electric pickup truck on the market, the R1T, with its unmistakable doe-eyed headlights. Rivian's boxy, athletic-looking pickup was about to beat Ford's F-150 Lighting to market by about eight months.

It was weeks away from an initial public offering that had the potential to make some people in the room millionaires. Rumors were swirling in the press that Rivian could begin trading at a valuation of around $60 billion, an extraordinary amount for a company that was unproven in the cutthroat US car market and unlikely to turn a profit for years. Rivian was burning through about $1 billion a quarter and, if all went well, would sell 25,000 vehicles in the coming year.[1] GM and Ford, on the other hand, would post $14 billion and $10 billion in operating *profit* respectively in 2021, and combined sold more than 10 million vehicles.

Yet that stock valuation would have put Rivian's market cap on a par with GM's and Ford's. Steady profits were good, but investors clearly were looking for a home run, and they had a model for one. Tesla already had shown them the magnitude of the EV growth opportunity: its market value was sitting above *$1 trillion*, nearly ten times GM and Ford combined.[2] The excitement around Rivian had earned the company an informal moniker: the Tesla of Trucks. As the team gathered that night, Rivian was arguably the hottest car company on the planet.

Scaringe was dressed in his standard uniform—a kind of tech bro meets ski bum look: dark jeans, trim-fitting, earth-colored button-down shirt, and blue-and-white trucker hat. Then in his late thirties, the young entrepreneur's appearance hews closely to Clark Kent: tall and square-jawed, with dark hair, black-rimmed glasses, and an easy smile, which, at this meeting in the customer engagement room, never left his face. He was there to toast his most loyal employees, some of whom had gutted out the startup life for a decade, joining him when Rivian was barely a design sketch. There are so many barriers to entry for a new car company that the mere fact Rivian had begun producing those electric trucks was remarkable in itself, an achievement impossible without the people in the room. The employees gathered with spouses and small children, a nod to the sacrifice that the families made during years of grueling long hours amid the uncertainty of an automotive startup. It was a rare congratulatory moment for Scaringe, who, while relentlessly positive and optimistic, was not much for reflection or pats on the back.

Day-to-day, Scaringe was affable and upbeat, but dogged, always onto the next thing. It wasn't uncommon for one of his direct reports to get a Friday night text, asking them to revise some PowerPoint slides or fuss with some other task that bled into their weekend after a seventy-hour week. At the gathering, the CEO offered some kind words for his most loyal people and gifted the first truck off the assembly line to his wife, Meagan, who, along with their three young kids, undoubtedly also sacrificed much.[3]

. . .

Ten years before this moment, Rick Wagoner—then about two years removed from his job as General Motors' CEO—got a note that an entrepreneur based in Florida, of all head-scratching places, wanted to talk to him about building a car company. Wagoner, a Duke University economics major and Harvard MBA, knew that any entrepreneur set on starting a car company—in the wake of a severe recession—was either hyper-ambitious or a little crazy, or both. It would have been easy for Wagoner to dismiss it outright, or just take a cursory spin through the business plan he received. But even if he wanted to, he couldn't. The slide deck was so clear and so thorough that Wagoner found himself continuing to flip through the pages. "I thought it was interesting. You could tell he was thoughtful and smart," Wagoner would say later. "I was impressed this guy was trying to do this."

RJ Scaringe grew up in Florida across the Indian River from the NASA launch pads at the Kennedy Space Center and Cape Canaveral.[4] His father, Bob, owned an engineering firm that worked on highly specialized heat-management systems, mainly for government agencies.[5] Scaringe had been obsessed with cars since he was a kid. He would memorize specs about engines and transmissions the way other kids would their favorite players' batting averages. Scaringe would sometimes peek under cars in parking lots to catch a glimpse of how their exhaust systems were designed. As a teen, he would store spare car parts around his bedroom—a windshield in his closet, a hood under the bed, engine parts on his desk. He once helped a neighbor restore an old Porsche.[6]

In high school, Scaringe was known to be sociable and smart, humble, but with an unusual intensity. He was a tenacious forward on his varsity basketball team, all hustle and grit. One summer, he worked three jobs, seven days a week, to make a down payment on a house that he'd rent for income. He was seventeen.[7]

By the time Scaringe reached out to Wagoner, he had a PhD in mechanical engineering from MIT, where he researched internal-combustion-engine efficiency at the school's Sloan Automotive Lab. It was named for Alfred P. Sloan, the legendary GM CEO known for building the company into the world's biggest corporation in the mid-twentieth century.

Scaringe finished his doctorate in 2009, a time of turmoil for the US auto industry due to the Great Recession and rising gas prices. GM was bleeding cash and on the verge of a government bailout. As part of the restructuring, GM would radically downsize, promising to close more than a dozen factories and shed around twenty thousand jobs—including Wagoner himself who, as a condition of the government's support, agreed to step down after a tumultuous nine-year tenure.

As Scaringe watched these giants desperately try to revitalize themselves, he saw an opportunity. A smaller, leaner company, he figured, could target niche vehicle categories with lower sales volumes. In the car business, a shorthand truth held that car companies needed to sell 200,000 of a car model to turn a profit. Scaringe disagreed with the conventional wisdom, though, and produced reams of his own, carefully laid-out analysis to persuade prospective investors. In his mid-twenties, fresh from MIT, Scaringe founded his company in unused warehouse space at his father's company in Florida.[8]

"We think we can do better than the current manufacturers," Scaringe told *Florida Today* in 2010.[9]

Scaringe called his company Avera, but Korean auto giant Hyundai Motor Co. felt that name was too similar to its Azera sedan. So Scaringe changed the name to Rivian, a play on the Indian River that served as a backdrop to his youth.[10]

Rivian eventually would pin its identity to three ideas: electric, adventure, and trucks. But at the start, there was none of that. Scaringe's original vision was for a hybrid-diesel sports car developed with a small team of designers. At the time, Toyota owned the hybrid game with its hit Prius. The technology uses a small lithium-ion battery that feeds power to an electric motor. In a traditional hybrid, the electric system generally doesn't propel the car on its own but helps take a load off the car's gasoline engine, saving fuel and in some cases boosting power.

Scaringe believed he had a novel idea to achieve much greater fuel economy by mating a hybrid system to a *diesel* engine, rather than gas. Diesel engines typically get much better highway gas mileage than gas engines—that's why long-haul trucks use diesel fuel. Scaringe estimated that a diesel-hybrid system could double the fuel efficiency of the Prius and other hybrids. But he also wanted to differentiate his car in another way: style. Instead of a Prius-like wedge-shaped, sensible hatchback, Scaringe's design team sculpted a sinewy sports car.

"We're going to cut out a big chunk of our market and go after one piece by making our car so desirable, so fun to drive and so good to look at that they're going to want it," he told *Florida Today* in 2010.[11]

Of course, he wasn't the only young entrepreneur with that idea. Tesla's Elon Musk was pinning his startup's hopes on the sporty Roadster coupe and, by 2012, already was in the market with a second car: the Model S, a large, modern sedan that was like nothing else on the market. Both Musk and Scaringe were brilliant and ambitious, but the similarities didn't go much beyond that. If Scaringe channeled the mild-mannered Clark Kent, Elon Musk was *Iron Man*'s Tony Stark, the ultra-rich brash playboy inventor. In media interviews—once Rivian got on the radar of the automotive press—Scaringe was careful, diplomatic, and stuck to precisely scripted talking points. Musk, eleven years older, was brash, provocative, irreverent—at times juvenile, both in his in-person comments and on Twitter, the social media site that he would one day acquire.

And yet, these two automotive interlopers were closely aligned in one major aspect of their strategies: they wanted to make desirable, sexy cars that also happened to be environmentally conscious. Traditional car executives still seemed to view green cars as a chore, something they needed to do to check a regulatory box. These two men saw that as an opportunity to fill a void.

. . .

Florida turned out to be a lousy place to start a car company. Scaringe naively figured there must be parts suppliers to the boating industry there he could tap. There weren't. Design and engineering talent was too scarce. To deal with that, Scaringe turned to Detroit for help. This was another difference with Musk, who had long chided the likes of GM and Ford for their old, lumbering ways. Scaringe loved the auto industry and sought out its luminaries for guidance, not ridicule. Many of the early Rivian employees were automotive designers recruited from Detroit.[12]

What impressed Wagoner was Scaringe's unconventional approach. He wasn't trying to follow the time-tested conventional pathways to profits— huge volume and low marginal costs. Wagoner also had heard that Tom Gale had joined the Rivian board, a former designer at Chrysler who had a hand in icons like the Lamborghini Diablo and Dodge Viper and whom Wagoner long admired.

Scaringe invited the former GM CEO to see his small shop. "Why don't you come by and see what we're doing?" he asked.

Wagoner was traveling to Florida for a meeting elsewhere and agreed to drive a few hours out of the way to Scaringe's little warehouse. He left impressed, again, and eventually agreed to invest some of his own money, and he joined Rivian's board.

Big investors, however, were hard to come by. Wagoner and Gale worked their respective networks, but they didn't provide a big break on funding. What did was Scaringe's dogged pursuit of MIT's alumni network. He was introduced to the Jameel family of Saudi Arabia, whose Abdul Latif Jameel, or ALJ, business empire was built by being the first Saudi distributor for Toyota.[13] Scaringe became particularly close with Hassan, the family's youngest scion.

Once, on a visit to Saudi Arabia, Jameel showed Scaringe a 561-foot flagpole that his family had helped construct in the city of Jeddah. The two young men spontaneously decided to climb it. Scaringe took off his jacket and tie, donned a climbing harness and safety helmet, and the men spent two hours clambering up the white metal pole under the searing desert sun.[14] ALJ eventually would invest hundreds of millions of dollars in Rivian. Its faith would be richly rewarded in the 2021 IPO.[15]

Scaringe's pivot to trucks makes sense in retrospect. For one, pickup trucks and big SUVs were among the steadiest selling and most lucrative vehicles in the US market. For another, the green-minded Scaringe wanted to target the sector's biggest polluters. And in conversations with the Jameels, Scaringe had begun considering rugged yet efficient pickups and SUVs that could withstand the harsh rigors of desert sand.

Wagoner also reinforced Scaringe's gravitation toward trucks. The former GM CEO knew better than just about anybody how profit rich the truck market could be. GM's Silverado along with Ford's F-150 and Chrysler's Dodge Ram, and offshoots of those models account for more than 90 percent of the market for big pickup trucks, as well as the SUVs built on the same truck platforms, giants like the Cadillac Escalade, Chevy Suburban, and Ford's Lincoln Navigator. For the Detroit companies, those models account for less than half of their vehicle sales, but the vast majority of profits.

Detroit's stranglehold on the market is rooted in an early 1960s trade tariff that imposed a 25 percent tax on imported trucks. The levy was dubbed the "chicken tax," because it was in retaliation for some European countries placing a tariff on chicken from the United States.[16] Eventually, Toyota and Nissan circumvented the tax by opening US truck plants. Toyota even built its pickup factory in San Antonio, the beating heart of truck country, hoping to persuade American-flag-flying pickup buyers. But it hasn't worked. Toyota's Tundra typically sells less than one-third what number-three Ram sells. Since the Great Recession, the Detroit companies have been offering a seemingly endless parade of new features and gadgets to push the price of pickups well into luxury territory. It's not uncommon for US pickup buyers to pay north of $70,000 for their rigs.

Soon after ditching the sports-car idea, Rivian toyed with the concept of a gas-powered truck. Later, a hybrid truck was on the table. But as Tesla's Roadster was winning rave reviews for its peppy performance, Scaringe decided that all of his company's vehicles would be electric.

Rivian went into "stealth mode," a term to describe a young company working out of the public's eye to develop its first product. At this point, 2015, the Florida problem could not be ignored any longer. Scaringe needed to line up suppliers and he needed more talent—engineers and managers adept in the byzantine world of supply chain management and automotive purchasing. Scaringe moved headquarters to Livonia, a Detroit suburb.[17]

Rivian took up residence in a small office in a nondescript two-story, brick-and-cinder-block building. The company kept a minimal web presence, a small site that focused on recruiting auto engineers, with vague language about reinventing mobility.

Next, Rivian needed to find a place to build the trucks that would reinvent mobility.

. . .

In 2016, Scaringe dispatched a small team to a mothballed factory formerly run by Mitsubishi Motors North America. The shop turned out nearly 5 million vehicles over thirty years, mostly cheap and sensible ones like the Mitsubishi Eclipse, Eagle Talon, and Plymouth Laser.[18] But Mitsubishi was reeling after the Great Recession and shut the factory's doors in late 2015. A Google Earth image taken in 2016 shows a paved lot next to it where someone has arranged large metal objects—maybe car frames, maybe containers—to read "THE END!" It can be seen from the sky, two miles up. The owners were having something of an estate sale, hoping to unload as much of the tooling and hulking assembly equipment as they could before the building would likely be demolished.

The team reported back to Scaringe: yes, there were hundreds of pieces of equipment tagged with fire-sale prices that Rivian might want to snag. But the massive, grimy, dimly lit factory itself was also appealing. What if this could be Rivian's future assembly plant? They'd need to dispatch more decision-makers on the five-and-a-half hour drive down to Normal to take a look.

Larry Parker was among the Rivian employees to make the second trek. A Detroit-area native with a youthful energy and a bent for storytelling, Parker was among Rivian's first few dozen employees. His father worked as a machinist at an auto company, and the younger Parker always assumed he'd follow in his dad's blue-collar footsteps. Parker was a hands-on car enthusiast; he'd built his own rally cars, including a souped-up Eagle Talon he concocted in his early twenties, a car that came off the assembly line in Normal. Parker eventually became interested in design and enrolled at Detroit's College for Creative Studies, a hotbed for future automotive designers. Despite his car roots, though, Parker had no interest in joining the auto industry—he viewed himself as an innovator and thought the car business was too rigid and set in its ways. He studied industrial design. His first jobs included working for aerospace companies and furniture makers in western Michigan.[19]

In 2015, on the suggestion of a friend, Parker showed up in Livonia, somewhat skeptical, but checking on available jobs. At one point he walked past a curious-looking, two-seat pickup truck, with the engine oddly tucked behind the seats, in a spot he knew was called the "mid-engine position." Parker took a minute to study the underside of the truck, which was mounted on a hoist.

"It was this weird, weird little concept car they had hand-built," Parker recalled. "But it was clear that whoever had been playing with this thing knew what they were doing. It had an advanced, multipoint suspension, this mid-engine layout. I was intrigued."

During his interview, Parker showed Scaringe a $600 trash can that he had designed for a previous client, for which he'd spent a lot of time in people's homes, observing how they use, well, their trash cans. Scaringe already had plenty of car designers on staff, but he figured Parker could help him develop Rivian's future brand direction—the aesthetic and ethos that would eventually evolve into its outdoorsy, sustainable, adventure identity.

Arriving at the idled Normal factory, Parker remembers, "It looked like people had literally just gotten up from their desks and never came back. Like there were coffee cups with coffee still in them." But there was also a giddy enthusiasm among the group as they rode through the plant on three-wheel bicycles that were left behind by Mitsubishi, used by workers to quickly traverse the massive grounds. On one hand, it was a much bigger facility than Rivian's production aspirations called for. It had a giant stamping press for spitting out whole side panels of sheet metal, a process that automakers typically do in an entirely separate facility.

At this point, Scaringe and his team barely knew what Rivian wanted to be. Yes, it had a prototype for a futuristic-looking truck. Scaringe back then viewed Rivian very much as a tech company focused on being part of the future of mobility. In 2016, prevailing winds suggested that the future of the industry was, yes, electric, but also not focused on ownership. The notion that people would be sharing their cars, maybe paying a monthly subscription for them, or getting them on demand, was hot. Uber and Lyft had exploded onto the scene. Autonomous technology had mesmerized investors and the media.[20] Scaringe still wasn't sure what transportation challenges his future Rivian would be trying to solve. But the idea of a more modern, tech-forward business model didn't really square with the old-world, move-the-metal focus of traditional automakers. For Rivian, maybe the future looked more like, say, British carmaker Lotus, a small-scale operation of boutique, hand-built vehicles that could be sold to a discerning customer base at high prices. At this point, nothing had been decided; all options were on the table.

Once Rivian's leadership saw the old Mitsubishi plant, though, their view began to change, and so did its truck design.

"I think that that factory really played a big role in how we thought about the vehicle," Parker says. "The plant completely changed the R1T."[21] Being able to take over such a large-scale operation, with capacity for more than 200,000 vehicles annually, opened the possibility of Rivian emerging as a potential mass-market player. The original design for the R1T was futuristic looking, a niche product. The opportunity to pick up the Normal plant got Scaringe and his team thinking about a more mainstream product, so its styling was toned down—still modern and different, but in a more conventional pickup-truck form that might draw in conventional truck buyers and sell in bigger volumes.

In late 2016, Scaringe closed the deal for the plant with a signing at the hip Coffee Hound café in the adjacent college town of Bloomington.[22] Rivian got the factory for a relative steal at $16 million, promising to invest another $175 million to bring the plant back to life and employ at least a thousand workers.[23] Here again, the Rivian team followed the template of Tesla, which rode the coattails of the auto industry by scooping up a defunct factory on the cheap in Fremont, California, that had once turned out Chevys and Toyotas.

Scaringe's team would also get its own crash course in what Elon Musk once described as "production hell."

. . .

Nearly two years later, Rivian was still a relatively well-kept secret, even though it had raised several hundred-million dollars. The company would remain in stealth mode a few months longer while Scaringe worked to get on the radar of a key person: Amazon CEO Jeff Bezos.

Scaringe had heard that Amazon was looking for an automotive partner so it could begin converting its burgeoning fleet of delivery trucks to electric. He worked some connections to broker a meeting with Bezos. To his surprise, Bezos agreed fly to Detroit to visit this little-known startup, which by then had moved into more-spacious digs in Plymouth, Michigan, an affluent suburb between Detroit and Ann Arbor. . Bezos and Scaringe met in a conference room, and RJ white-boarded his vision. Bezos left impressed and instructed his team to explore a potential contract for future electric delivery trucks—and a possible equity stake in Rivian, which was then valued at less than $2 billion.[24]

It's hard to stay in stealth mode with that sort of unicorn valuation and an investment interest from Jeff Bezos. Rumblings of the company's ambitions had started to spread in tight-knit automotive circles around metro Detroit. The new headquarters was a short drive from the home of one of Ford's top executives: Joe Hinrichs, a Wagoner mentee. Partnerships between tech companies and traditional automakers were en vogue at the time, and Wagoner thought Hinrichs might want to take a peek at Rivian's operation. He called Hinrichs and told him, "I know you and RJ both really well. I think you'll hit it off."[25]

Wagoner was pleased to have potentially lined up Rivian with a heavyweight investor and technical partner. But, while driving his car later that day, a thought hit him: had he just boxed out GM of a potential opportunity by connecting Scaringe to Hinrichs? Wagoner later called then-GM president Mark Reuss, Mary Barra's top lieutenant, to tell him about Rivian and brokered an intro to Scaringe for GM as well.[26] Wagoner didn't know it at the time, but he had just set off an epic game of brinksmanship between the archrival automakers.

An Ohio native, Ford's Hinrichs broke into the auto industry as a supervisor at a GM factory in Dayton. He joined crosstown rival Ford as a plant manager in 2000 and rose the ranks to hold executive posts across the company. He was credited with jumpstarting Ford's growth in China in the early 2010s. Later he championed the return of the iconic Bronco SUV. He clashed with the Trump administration on fuel-economy regulations. About a decade and a half older than Scaringe, Hinrichs was well liked inside Ford and comfortable at the bargaining table, having previously led contract talks with the United Auto Workers.

Hinrichs swung by Rivian's headquarters, a twenty-five-minute drive from Ford's Dearborn campus and five minutes from his house. Scaringe again scribbled on a whiteboard to outline Rivian's history and strategy, going all the way back to its 2009 founding—the idea for the sports car, the outdoorsy Patagonia vibe.

The young CEO also explained that he was seeking an automotive partner to jointly develop a new vehicle using Rivian's EV platform. Scaringe figured that sort of deal would validate Rivian's technology and, at the same time, help his team learn how to launch a new vehicle, a process fraught with risk and pitfalls. Scaringe knew the steep manufacturing curve Rivian faced: figuring out how to precisely stamp huge slabs of sheet metal, for example, or put in place exacting quality controls. "RJ was humble enough to know what he didn't know," Hinrichs would say later.

"He knew how hard it would be to develop and build vehicles at scale."
One passing detail gave Hinrichs pause, though: Scaringe subtly hinted
that GM and Toyota might be kicking the tires on the startup, too.[27]

Later the two men toured Rivian's design studio, strolling past clay
models of future trucks. Hinrichs bumped into a few former Ford designers. The Ford exec left impressed, as most people seemed to, with Scaringe
and his grasp of the task at hand, his confidence, and Rivian's business
plan and the product designs.

The next day he returned to Rivian with Hau Thai-Tang, one of his direct reports and effectively Ford's top engineer in charge of all product
development. "I'm going to let you do all the talking," Hinrichs told Thai-
Tang on the way over.

Thai-Tang spent a few hours with Scaringe discussing Rivian's strategy for electric motors, batteries, and other technical matters. Scaringe
led them into a garage area, where there was, of all things, an F-150 sus-
pended on a hoist. The truck had been stripped of its internal-combustion
guts and in its place was installed Rivian's EV skateboard: a large battery
pack stuffed with hundreds of individual lithium-ion cells; electric motors
set at each of the wheels; and an inverter, which controls the speed of the
motors.[28] In those early days, the design of Rivian's truck was still veiled
in secrecy, so it used the Frankenstein F-150—called a test mule—so engineers could put Rivian's mechanical system through the paces undetected
on metro Detroit roads.[29]

"There really is something here," Thai-Tang told Hinrichs on the drive
back. Hinrichs knew the two of them had enough clout to get buy-in from
Ford's leadership. Within a few weeks, a bigger technical team of Ford
engineers was sent over to Plymouth for a deeper dive. They spent a few
days with Rivian, feeling out whether Ford could share either technology
or suppliers and learn a bit more about Rivian's tech setup. Ultimately,
though, they were trying to determine whether Rivan had the capability
to engineer and build the vision Scaringe had laid out. In those early days,
Ford had scouted more than a few small EV companies and tech firms—
the conclusion, usually, was that they were in over their heads, capable of
putting on a good show but, when pressed, incapable of delivering.

The team returned to Dearborn with a verdict. No, this newcomer didn't have any in-house tech that couldn't be found inside Ford, the proud engineers reported. *But*, in the end, they concluded, Rivian was legit. Its capability wasn't "vaporware," the term used for startups that try to peddle a concept without any real tech or product behind it.

Ideally, then, Ford would strike a deal that would give it a closer look at Rivian's tech and use its platform to do at least one future electric model. Ford's CEO at the time, Jim Hackett, told Hinrichs that he shared his interest in doing something with Rivian, but Ford's finance team was dissuading him from it. Ford had been missing its earnings targets and was under pressure from Wall Street to boost cash flow.

Hinrichs reluctantly told Scaringe that a deal looked unlikely in the near term. But he said to stay in touch. "Things can change quickly," Hinrichs told him.

. . .

The week before Thanksgiving in 2018, still with no deal in place with Ford, Scaringe strutted onto a stage at the iconic Griffith Observatory in Los Angeles, on the eve of the LA auto show. LA's show is comparable size to the big shows in New York and Chicago, but it holds special status as the go-to venue for green-vehicle debuts. LA is EV country: nearly half of all battery-powered cars in the nation are sold in California.[30] It wouldn't be long before Teslas roamed LA's famously clogged highway system in numbers near to the ubiquitous Honda Civic.

By this point, there was enough chatter about the stealthy Rivian that the audience around the stage was packed dozens deep, including pop superstar Rihanna. Scaringe was dressed in his trademark dark jeans and button-down shirt, dark outdoorsy jacket, and tan shoes. He looked ready for a day of rock climbing. The backdrop was a massive video screen playing footage of Rivian trucks, first in clay model form, then in their sheet-metal flesh, tearing through the forest or along the beach. Dramatic music pulsed through the auditorium as a shiny silver R1T, the pickup, and a pine-green R1S, the SUV, rolled onto the stage, drawing whistles and cheers.

Scaringe boasted that the truck could rip from zero to sixty miles per hour in about three seconds. The oval-shaped headlamps, like wide-open, friendly eyes, were designed to be instantly recognizable. The materials used inside the cabin are premium but durable, like a ski jacket, he said. The front trunk, or "frunk," could fit a cooler and a big backpack. Many other EVs have frunks, but Scaringe said these vehicles could do one better: a large storage tunnel running from one side of the truck to the other, behind the rear seat, which could fit skis, fishing poles, even a surfboard. The tunnel opening flips down as a seat that the user can sit on to gear up, he said. "Let's say you're going to the beach and putting on a wetsuit," he said, or "putting on your boots before a big hike." The examples were deeply deliberate: sure it's electric, but you can beat the hell out of it through all of your outdoorsy pursuits.[31]

The Rivian vehicles landed on just about every "best of LA auto show" list that year. "You wouldn't think you could reinvent the pickup. But that's exactly what Rivian has done," *MotorTrend* declared.[32] Like most Rivian employees on the day of the reveal, Larry Parker was feeling reflective. Years of long hours and stress were paying off. Rivian trucks were stealing the show amid giants of the auto industry. He sent a heartfelt text to Scaringe later that day, congratulating him on the huge milestone.

"They don't like the headlights," the CEO shot back. Scaringe was already on to the next thing.

. . .

With Ford talks on the back burner, Scaringe and his team continued discussions with a group of investors led by Amazon. Meanwhile, GM executives had gone through a similar process as Ford with Rivian, but in this case the deal that started to take shape was more far-reaching. GM wanted to do more vehicles and eventually integrate Rivian's technology as a platform for future GM vehicles. Scaringe had hit it off with Mary Barra and

Mark Reuss. And clearly GM had the deep pockets to do a substantial deal. Even so, the scope of the proposal made Scaringe a little uneasy. He worried that future GM projects would bog down his team and steal resources from the R1T and R1S, which were his true passion.

Scaringe's team paused talks with GM to finalize the Amazon deal, which was announced a couple of months after the successful debut at Griffith Observatory. The tech-and-retail giant would lead a group of investors in a $700 million infusion into the EV truck maker, making Amazon Rivian's largest minority investor.[33] The deal had the potential to raise Rivian's profile and rocket up its valuation.

Even before that announcement, rumblings of Amazon's flirtation had renewed the interest from Ford's leadership team. Hinrichs had continued checking in with Scaringe, who deftly signaled the imminent Amazon deal and dropped hints about conversations with Barra's team at GM. During dinner at a high-end Detroit-area steakhouse, Scaringe finally told Hinrichs flat out that Rivian had agreed to hold exclusive thirty-day talks with GM. Hinrichs was stunned.

In the car business, there are few battles as intense and rich with history as GM versus Ford in trucks. Maybe BMW versus Mercedes, or Ferrari versus Lamborghini, or the Ford Mustang versus Chevy's Corvette. But in terms of the fervency of their customer bases, the central role that pickup trucks play in their owners' lives, and the sheer magnitude of the profits at stake, the competitive zeal between Ford and GM on trucks was like none other, and it drew out the sharpest elbows.

The companies have spent untold billions in what amounts to a multimedia pissing match over whose truck is better, more powerful, longest lasting, and most capable of hauling and towing stuff. Occasionally things have gotten so heated that lawyers entered the scrap. In 2012, Ford threatened legal action after GM ran a Super Bowl ad that showed its Chevy Silverado surviving the Mayan apocalypse, while one guy who drove an F-150 presumably perished.[34] A few years later, after Ford switched the F-150 body from sheet metal to aluminum to reduce weight, GM ran commercials showing how a dropped construction tool or a load of cinder blocks could puncture the F-150's bed. Ford threatened to sue.[35]

Neither GM nor Ford viewed Rivian as a savior to usher them into the EV age. Both companies were in the early stages of their own electric truck projects. But they'd both been burned by not taking Tesla seriously. And now here was the "Tesla of Trucks"—trucks, Detroit's main profit driver— sitting right in their own backyards. The stakes were high enough.

Hinrichs listened to Scaringe while trying to process the very real prospect of losing this deal to his archrival. But Hinrichs also sensed an opening. Scaringe confided that the high-level talks with Barra and Reuss had gone great, but things got more tense during the thirty-day exclusive period, as his team tried hammering out details with GM's lawyers and other mid-level managers. GM's terms got more onerous. Scaringe's feelings of unease about GM wanting more than he was prepared to give intensified. That was just enough daylight for Hinrichs, who urged the young executive to keep an open mind.

"Just because you got engaged," he told Scaringe, a vegan in a steakhouse, picking away at some greens, "doesn't mean you need to get married."

The steakhouse conversation happened with only about a week remaining in the thirty-day exclusive period. As it happened, both Scaringe and Hinrichs were scheduled to be out in Seattle the following week, Scaringe to meet with Amazon, and Hinrichs to stop by Microsoft. Hinrichs made a proposal: "Since we're both going to be in Seattle, if you don't sign anything, why don't you come by my hotel to discuss things? I'll have a plane—you can fly back with us."[36]

At this point, things with GM seemed pretty baked. Bloomberg had reported that a deal between Rivian and GM was close.[37] The companies had already applied for federal antitrust approval, and their PR teams were preparing announcements.

Still, the thirty-day exclusive period came and went without a deal. Scaringe met Hinrichs in the lobby of a downtown Seattle hotel. As they sipped on Diet Cokes and water—neither of them drinks alcohol—they discussed the broad outlines of a workable deal.

The next day, they boarded Ford's Gulfstream for the flight back to Detroit. They sat up front until the plane reached cruising altitude, before moving to the rear of the aircraft and shutting the door. A handful of executives from Hinrichs's team remained toward the front of the plane, sworn to secrecy.

Any agreement, the executives decided, would have to be finished quickly. They spent the four-and-a-half-hour flight texting and emailing their teams as they worked on terms.

During the flight, Scaringe received a text from Barra, who had been traveling with Reuss. Rivian's legal team had notified GM that the exclusive period had ended without a finalized deal, and Barra asked Scaringe if he could talk when they both returned.

The Ford plane landed at the corporate-jet terminal in suburban Detroit in the evening. As the Gulfstream taxied toward its hangar, Scaringe and Hinrichs looked out their windows to see GM's jet parked just outside the hangar (the companies shared the space), with two black Cadillac Escalades pulled up beside it. Scaringe panicked. What if Barra saw him?

Hinrichs had the pilot phone ahead and ask that his Lincoln Navigator pull alongside the Ford plane. When the door opened, Scaringe dashed out of the plane and into the Navigator, keeping his head down as Hinrichs loaded their luggage. They would continue negotiations that night.[38]

A deal was finalized around 4 a.m. Ford would invest $500 million in Rivian and agreed to develop a Lincoln SUV using Rivian's technology. Hinrichs would get a seat on Rivian's board.[39]

When the deal was announced, Barra was upset. She didn't think Scaringe had handled the situation professionally. Her team huddled, determined not to let this snub stop them from hitting the EV truck market. Within days, they emerged from a closed-door meeting with their answer: a $113,000 Hummer super truck.

8

Toyota's Turn

By the end of 2022, Rivian had built more than 25,000 EV trucks. New electric models like Ford's F-150 Lightning, Hyundai's industrial-chic Ioniq 5, and Lucid's sleek Air sedan boasted waitlists stretching months or even years.[1] Carmakers fortunate enough to have electric models to sell were greeted with zealous buyers willing to pay several thousand dollars above asking price. US EV sales surged nearly 70 percent for the year, compared with a decline for non-EVs.[2] Buyers were paying in the mid-$60,000s on average, some $15,000 more than the industry average. Ask an auto executive around this time about their EVs, and they would have lit up.

But not Jack Hollis of Toyota. At a December 2022 media event, Hollis sat on an upholstered chair inside—of all places—a velodrome tucked in a bustling area north of downtown Detroit that was sponsored by the Japanese automaker's luxury Lexus brand. Hollis worked out of Toyota's US HQ in Texas but was in town with other top brass for an annual holiday party and year-in-review chat with the automotive press. As cyclists in helmets and tight-fitting Lycra zipped around the steeply embanked track, Hollis, then Toyota's top US sales executive, settled in for a slate of interviews.

There was plenty to cover. A still-out-of-whack supply chain, a vestige of the pandemic, meant his US dealers had very few cars on their lots. Eager buyers snapped up everything the company built as soon as it rolled off a truck at the dealership. Toyota's net profit for the year would blow past analysts' forecasts.

A California native with a six-foot-two athletic frame, Hollis played outfield on a national championship Stanford baseball team in the late 1980s and was drafted by the St. Louis Cardinals. Now in his mid-fifties, with spiky silver hair, Hollis is an effusive, fast talker, full of positive energy. He has spent his whole career at Toyota, moving up the executive ranks in US sales and marketing. Hollis raved about red-hot demand for a new Toyota Sienna minivan and how customers were flocking to Toyota's growing portfolio of hybrids.

Then Hollis was asked about the Toyota brand's lone fully electric car on the US market, the bZ4X. The midsize SUV with the unwieldy alphanumeric name had received decent feedback from car critics. A shade bigger than Toyota's top seller, the RAV4 SUV, the bZ4X was, of course, more expensive, starting in the mid-$40,000s, with an electric range of around 240 miles. Like most EVs, it was zippy and fun to drive, though not as quick as many competitors, and had an upscale interior. It was a fine first foray into EVs for Toyota.[3]

There was just one problem: the company seemed almost disinterested in selling the thing. Toyota would deliver only 1,220 of them in the United States during 2022—about the same number of RAV4s it sells on a decent Saturday.

Hollis leaned back in his chair at the question. The feel-good California vibes faded. He answered that, yes, there is enthusiasm around the bZ4X. Then he all but waved off the topic. "I mean, it's tiny volume," he said, quickly redirecting the conversation.[4]

You might think by this point that Hollis would want to lean into Toyota's EV story. Some of the new electric models from rivals were getting traction in the heart of Toyota's most important markets, like California. Ford Mustang Mach-E was a hit there. A Rivian driver in San Francisco or LA in 2022 was likely to be swarmed by curious bystanders. Tesla continued to chew into Toyota's market share in Cali with the Model Y and Model 3, the two top-selling models in the state, edging out the RAV4 and Camry sedan.[5]

But Hollis wanted to drive home a point: despite the EV hype, much more was needed to solve the real problem of reducing planet-warming greenhouse-gas emissions, he said. You need to bring an arsenal of different weapons, Hollis insisted: hybrid cars like the Prius, which get great gas mileage but generally don't operate in full electric mode. Plug-in hybrids, which do offer all-electric driving for a certain number of miles, before switching over to gas. Toyota was even experimenting at the time with an internal-combustion engine that burned hydrogen instead of fossil fuel, emitting very little carbon and other pollutants.

"Why are we so focused on one answer, when if we used multiple answers, we'd actually beat carbon faster?" Hollis asked. He had reason to feel defensive when pressed on Toyota's EV strategy. The Japanese juggernaut had long enjoyed a reputation as the darling car company among environmentalists. Now, that reputation was under attack.

Toyota has dominated the market for hybrid cars since introducing the Prius in Japan in 1997. It wasn't anything special to look at, but under the hood, it was a revolution: a 1.5-liter, four-cylinder gas engine mounted parallel to an electric motor that was hooked to a small, nickel-metal hydride battery. Neither of those power trains was enough by itself, but combined, the result was a triumph of fuel efficiency. The Prius was pokey in a straight line, clocking "a glacial 13-second run to 60 mph," *Car and Driver* magazine wrote years later. But it was peppy in stop-and-go traffic and, most importantly, delivered an eye-popping fifty-two miles-per-gallon average.[6]

When the Prius made its way to the US market in the early 2000s, the $20,000 car quickly became a favorite of the environmentally minded. It also became an unlikely status symbol for celebrities. Owners included Billy Joel, Leonardo DiCaprio, and Seinfeld creator Larry David, who was said to have bought three of them, in part to please his wife at the time, Laurie, an environmentalist.[7]

The Prius became a global hit. By 2010, its worldwide sales topped 2 million. Toyota then estimated that, since the Prius's introduction, it had saved 11 million tons of carbon from entering the atmosphere.[8] The stubby hatchback cemented Toyota's sterling-green image and paved the way for broader interest in electric cars by introducing the concept that adding electrons to the power train could not only save fuel, but enhance the driving experience with responsive, quiet acceleration.

Starting around 2017, though, when GM, VW, and other competitors began making a sharp turn toward fully electric cars, Toyota didn't follow. Executives assured investors that they were in the EV hunt, but that promise also came with some mixed messaging that led critics to claim that Toyota's approach to EVs was half-hearted.

Environmental activists already had been scrutinizing Toyota's public statements and advertising for anti-EV undertones. Several times, the automaker was chastised for maligning fully electric cars, portraying them as a pain to charge. At times, its ads described Toyota or Lexus hybrids as "self-charging," a silly description meant to make a jab at EVs' need to be plugged in. Other ads described their hybrids as having "infinite range," which, critics pointed out, would imply that there's no need to stop at a gas station.[9]

In 2019, the EV-focused news site Green Car Reports posted a story that took Toyota to task for a commercial it aired in the UK featuring a hybrid version of its Corolla. The spot showed the car zooming past an electric vehicle sitting idle as it charged by the side of the road. "It's disappointing to see Toyota falling back on that old trope that charging is inconvenient and ineffective," wrote author Eric C. Evarts. "That market of millions of buyers who want to do right by the environment have already begun looking elsewhere."[10]

Soon, the critics wouldn't need to scour regional marketing campaigns for cryptic signals about Toyota's ambivalence to EVs. They could simply listen to Toyota scion and CEO Akio Toyoda.

Toyoda took over his family's company in 2009. Then in his mid-fifties, the affable, energetic executive was not the stereotypical modest, buttoned-up, gray-suit Japanese corporate leader. He is the grandson of Toyota's founder, Kiichiro Toyoda, who in the 1930s, convinced the Toyoda Automatic Loom Works—a maker of machines used in the textiles business—to start a car division. The automaker changed the "D" to a "T" to become Toyota, in part because it took eight strokes to write, and eight is considered by some a lucky number in Japan.[11] Akio's father, Shoichiro Toyoda, ran the company for a decade, until the early 1990s, a time when the automaker became known for its manufacturing precision, top-notch quality and reliability, and great fuel economy.

Akio Toyoda spent the first several years of his tenure navigating a string of crises. The global economy still was in the throes of the Great Recession. Toyota was mired in a years-long safety scandal involving cases of unintended acceleration in its cars. A 2011 earthquake and tsunami in Japan would scramble global supply chains.

Once through those challenges, Toyoda's fun-loving persona began to shine through. He would show up at racetracks around the world, clad head to toe in Toyota-red racing gear, ready to hop behind the wheel and compete.[12] He emphasized returning the company to its roots of "fun to drive" cars and more compelling looks, seeking to roll back Toyota's reputation for boring design. One of the first new models under his watch was the GT86, a sporty, fastback coupe.

As his biggest global rivals proclaimed their intention to be leaders on EVs, Toyoda was making it clear that he wasn't worried about pole position for that race. He told investors that EVs were just one item on a broader menu of power-train options.

During a 2018 speech at the annual consumer-electronics show in Las Vegas, he announced that Toyota would have at least ten fully electric models on offer by the early 2020s. But he simultaneously signaled his hesitancy. "I'm less concerned about getting there first as I am about getting it right, and about finding ways to use technology to benefit as many as possible."[13] The subtext: for a vast majority of Toyota's customers, EVs simply weren't the best alternative.

It wasn't an unreasonable view. Toyota is arguably the most global of any big car company. It's a major player in each of the three biggest regions: North America, Europe, and China. It's also more deeply entrenched than competitors in developing markets too, from Central America and Southeast Asia to Africa, where its rugged Hilux pickup trucks are ubiquitous. In many of those places, local power grids are unreliable even to power homes. For Toyota, going all in on electrics would mean leaving behind millions of customers for whom routinely plugging in their cars would be impossible.[14]

As the drumbeat of automaker proclamations about going all-electric grew louder, Toyota's PR machine went into overdrive to defend its market-leading hybrid technology. Realistically, internal-combustion cars would stick around for decades to come, the company reasoned. So what was wrong with getting 35 percent better fuel economy on a gas-engine car by adding a hybrid system that didn't require any plugging in? And are we sure EVs are that much better for the environment? Toyota told its dealers that the minerals needed to assemble the battery for one electric car was enough to make ninety hybrid vehicles, a claim backed by some experts.[15]

But in the early 2020s, the subject of EVs had crept into the culture wars—and the middle ground Toyota had staked was an unenviable position. It left Toyota open to attack from both sides. The campaign to combat climate change had the most vocal environmentalists espousing a full embrace of electrics, and critics eviscerated Toyota's pragmatic approach.

"[Toyota] has become the industry's biggest obstacle to the all-out transition to electric vehicles," Greenpeace said in a 2021 report, which ranked Toyota dead last among ten major automakers for having a plan to transition away from fossil fuels.[16]

Jack Hollis was firmly with Akio Toyoda in the middle ground. At the velodrome in Detroit, despite efforts to redirect the conversation, he did come back to the subject and talked extensively about all the complications, the alternatives, and the need for options beyond just electric. He'd gone on for so long about all the other ways in which automakers could fight carbon, his PR handler finally chimed in. "I mean, we *do* have some EVs coming," he said with a sheepish smile.

. . .

Of all those customers who began buying Priuses in the United States in the early 2000s, Margo Oge was among the earliest. She had a BMW 5 series sedan, purchased by her husband, but it mostly sat idle in their Washington, DC–area driveway. She couldn't bring herself to drive it often because her conscience was gnawing at her. Instead, Oge went out and bought a lime-green Prius.

Oge's job had something to do with that. She was director of the Environmental Protection Agency's Office of Transportation and Air Quality, a post she would hold for nearly twenty years, through 2012. That title made her among the most influential people on the planet when it comes to the type and amount of vehicle emissions that can spew from the tens of millions of cars, trucks, and buses.[17]

Oge, a petite woman with sandy blonde hair, grew up in Athens, Greece, and lived there with her family until moving to the United States as a teenager. Her early years in the bustling, ancient metropolis influenced Oge's path toward environmentalism. She remembers the constant air pollution. She recalls seeing the Parthenon's ancient marble pillars being coated in soot, and reading articles about how the pollution threatened the future of the monument.[18]

Once in America, Oge got an engineering degree at the University of Massachusetts Lowell. One of her first jobs was on the staff of Republican Rhode Island senator John Chafee, who was chair of an environmental committee. Oge drafted a bill that made plastic rings for beverage six-packs biodegradable to reduce the risk to marine life. She was hooked and began a long career of environmental policy and advocacy.

Decades later, despite her husband's best intentions with the BMW, Oge says she "felt like I needed to walk the talk." And, since there were no EVs on the US market at the time, that meant owning a Prius. She would dutifully climb into the car each morning for her commute along the George Washington Parkway in Arlington, Virginia, and grip the wheel tightly. "It felt very flimsy. And it was ugly. Sort of, cute-ugly," Oge recalled later, in her retained Greek accent. "I thought, 'OK, I'm doing a lot for the environment, but I hope I won't die if somebody hits me.'"[19]

There are two federal entities ultimately in charge of regulating the impact that the US vehicle fleet has on the environment. One is the EPA, where Oge worked: the Clean Air Act of 1970 gave it the authority to set limits on the pollutants released from vehicle tailpipes, like nitrogen oxides or volatile organics. A separate entity, the National Highway Traffic

Safety Administration, determined what kind of fuel efficiency cars and trucks sold in the United States should have. NHTSA was created in the 1970s, during the oil-embargo crisis, tasked with improving vehicle safety but also aimed at devising rules that would lessen the nation's dependence on foreign oil.

But by the early 2000s, as concerns were building over global warming, environmentalists were asking: what about carbon dioxide? That remained a blind spot—there was no federal agency overseeing carbon and other greenhouse-gas emissions, no US laws restricting how much of those planet-warming materials could be released from tailpipes. Environmental groups petitioned the EPA to treat those greenhouse-gas emissions as pollutants. California at the time also was drawing up rules to curb greenhouse gases. The issue wound up in the courts for years until 2007, when the US Supreme Court ruled that carbon emissions were indeed a pollutant under the Clean Air Act, because they affected the environment and public health. For the first time, it was fair game for the federal government to regulate them.

A massive effort to draw up the toughest rules for tailpipe emissions in the nation's history followed. Oge's team at the EPA joined NHTSA and California regulators and spent years wrangling with the auto industry's powerful lobbyists to hammer out rules that would govern how much fuel America's cars, SUVs, and trucks could use over the span of more than a decade. In the end, the headline number was jarring: car companies would need to hit 54.4 miles per gallon on average by the 2025 model year, basically double the 27.5 miles per gallon standard they were expected to hit in the early 2010s. [20]

All of the automakers came to the table with their own personalities and priorities. GM and Ford, for example, wanted to make sure the rules weren't as stringent for trucks and SUVs, terms they ultimately won. For Oge, Toyota was among the most cooperative carmakers. She had visited the company's headquarters in Tokyo and seen its vast engineering resources. Toyota brass would visit her in DC often. And, of course, Toyota was as well positioned as any car company to find success with these new, more-stringent standards. It had the Prius and, overall, among the most fuel-efficient lineup. Toyota's average fuel economy was 25.4 miles per gallon at the time for its 2009 model year. GM, Ford, and Chrysler were near the bottom, around twenty miles per gallon.[21]

"Toyota was always the first to raise their hands and say, 'We can do it,'" Oge recalled years later. "And the domestics were always lagging behind."

The new fuel-economy rules did not trigger a pivot toward electric cars. Instead, the focus for much of the following decade was on tricks and hacks to squeeze every drop of efficiency from existing internal-combustion engines, a race of sorts between engineering teams at the different companies. A common tactic to steal miles per gallon back was to put vehicles on a diet. Just as a lighter person burns fewer calories on a run or bike ride than a heavier one, cutting the weight of a vehicle improved fuel efficiency. Light-weighting became a mantra across the industry in the 2010s. Mercedes in 2012 switched to aluminum from steel sheet metal for its SL roadster, the brand's first-ever use of the lighter metal for the body of a production car. GM began dabbling with carbon-fiber pickup-truck beds. Ford spent a few billion dollars to switch its F-150 pickup truck—the company's crown jewel and profit engine—to an aluminum body, which shed about seven hundred pounds from the steel sheet metal it had used for decades.[22]

From the companies' perspectives, begging, borrowing, and stealing miles per gallon on an internal-combustion engine was a more practical way of hitting the tougher fuel-economy targets than selling EVs, which would hit the targets more easily. But developing new electric models would require brand-new vehicle platforms and a massive reshuffling of supply chains. Those would take years and tens of billions of dollars. So the car companies did little with EVs despite the shocking target they needed to hit.

Still, there was some dabbling—mostly thanks to California. Because of the state's infamous smog and air-pollution problems, the Clean Air Act granted California the right to set its own environmental regulations. The state had been hounding the car industry to move toward electrics since the early 1990s, when it adopted rules that eventually required that a small percentage of each automaker's sales in the state be electric. California regulators sparred with the industry over its so-called Zero Emissions Vehicles or ZEV program, the terms of which were delayed and watered down for many years.

But in January 2012, California got serious: electrics as a percentage of a carmaker's overall sales in the state would need to account for 10 percent of their total sales by 2025. Companies that didn't meet the targets would need to buy credits from competitors that exceeded the minimum thresholds.[23] Those credits would become a vital source of income for Tesla as traditional automakers dragged their feet on putting out battery-powered vehicles. Over more than a decade, through 2023, Tesla racked up around $9 billion in revenue from selling carbon credits to rivals.[24]

California's rules prodded the industry to produce a handful of EVs. Nissan's Leaf and GM's Chevy Volt in 2010 were among the first and set off a string of other electrics and plug-in hybrids. Volkswagen made a battery-powered version of its popular Golf hatchback. Ford did the same with its Focus small car. All of these models earned the derisive moniker "compliance cars." The companies sold the bare minimum to meet the regs, and every sale was a major money loser. GM at one point was losing roughly $10,000 on every Volt it produced.[25]

Even Toyota came out with an all-electric entry for the US market: a battery-powered RAV4 that traveled about eighty miles on a single charge. But company execs didn't put a lot of effort into the thing. Most of its drivetrain was provided by Tesla, including the battery and electric motors.[26] The companies had a partnership dating back to 2010, when Toyota invested $50 million in Elon Musk's startup and also sold Tesla the mothballed factory in Fremont, California at a discount. Toyota sold its stake in Tesla in 2017.

Toyota's lukewarm view of EVs was grating for another influential environmental regulator: Mary Nichols. A feisty lawyer with cropped salt-and-pepper hair, Nichols for many years ran the powerful California Air Resources Board (CARB), the group in charge of enforcing those EV mandates. Because California's environmental policies were adopted by about one-third of US states, CARB was considered as influential as the EPA itself and made Nichols one of the nation's most prominent regulators.

"When it came to the question of electric vehicles, Toyota's position was just: 'No, we aren't interested. We don't believe it. We don't think the public will buy them,'" Nichols recalled later.[27]

Instead, Toyota built out its lineup with more hybrids, including hybrid RAV-4s, which eventually outsold the Prius. This worked for Toyota, which continued to enjoy environmental cred. It still frequently appeared on the list of the best global "green" brands, tracked by consultancies like Interbrand and Deloitte.

But in 2019, Toyota's EV apathy would come to a head, and a perceived betrayal of environmentalists incensed some of the groups that had for so long sung its praises.

It turns out that the Trump administration didn't like those stringent fuel-economy rules Oge helped write a decade earlier and proposed a big rollback. If left in place, the emissions regulations would raise the prices of cars for American consumers, Trump's administration concluded. Many of the major car companies were quietly cheering Trump on. Auto executives were upset that the Obama administration had stuck with the rules even though they had become unrealistic, based on American's continued preference for larger, thirstier vehicles.

Trump's EPA proposed a significant watering down of the standards, requiring carmakers to improve fuel efficiency by only 1.5 percent each year from 2021 to 2026, rather than the 5 percent called for in the previous Obama-era regs.[28] The Trump administration also decided to take on California, moving to strip the state's decades-long authority to set its own environmental policy, including the EV mandate. The issue wound up in court.

But these counterpunches against the nation's environmental regulatory regimes went too far even for some auto executives. Yes, it would be nice to be off the hook from rules that they considered too stringent and unrealistic. And the car companies generally never liked California's special regulatory status. But they also had corporate reputations to uphold. And around this time, it was especially in vogue for big companies to proudly tout all the things they were doing to help combat climate change.

Plus, automakers already were developing more fuel-efficient vehicles for China, Europe, and other markets where rules were more stringent. They also worried that the feud between Trump and California could result in years of regulatory limbo, which would complicate their portfolio planning.

Suddenly everything was upside down: some carmakers were pressing the government for *stricter* regulation on tailpipe emissions. Most notably among them, Bill Ford. Then in his early-sixties, with bright blue eyes and sandy brown hair, he had long considered himself an ardent environmentalist. He had been pushing green initiatives at Ford since before it was cool, including introducing the industry's first hybrid-electric SUV in the early 2000s, and putting a sustainable roof of flowers and shrubbery on Ford's historic Rouge factory near company headquarters.

In the spring of 2019, he dialed the White House from his office in Dearborn, Michigan, to urge Trump to work with California. Ford explained that a big rollback in rules wouldn't help car companies, because they were all developing future models to meet tougher requirements in China and Europe anyway.[29]

Trump was confused. He thought easier fuel-economy rules would help the industry to sell more pricey trucks and SUVs. He told Ford that he was out of step with his competitors like GM and Fiat Chrysler. "He basically said [to Ford]: 'You're on your own,'" a person with knowledge of the call said.[30]

Bill Ford wasn't deterred. In the spring of 2019, he had Ford executives approach California's Nichols with a proposal: the company would voluntarily comply with fuel-economy standards that were tougher than the watered-down Trump rules, but a bit more lenient than the Obama-era ones that the president had quashed. BMW, Honda, and Volkswagen joined Ford.

This was a major fission. Car companies generally stuck together on fuel-efficiency policy. GM, Toyota, and other automakers not only refused to join the voluntary effort but made it known that, with Trump shielding them, they wanted no more of California's freelancing environmental policy. They backed the Trump administration's bid to strip the state of its right to have its own rules.

It shocked almost nobody that the Trump administration would take on blue-state California over emissions rules. But Toyota's backing stunned and angered the environmental groups and regulators that once saw the company as an ally. And it brought even sharper scrutiny of Toyota's stance on EVs.

Even Margo Oge, who while at EPA had worked so cooperatively with Toyota, rolled her eyes at what she increasingly saw as the company's foot-dragging on EVs. "Toyota's position hadn't changed over the years: 'It's too early. It's too expensive. We don't have the infrastructure. Hybrids are the way to get there,'" she said. "My pushback was always: 'But we're running out of time.'"

. . .

Akio Toyoda had become increasingly vocal about his concerns that EVs were being overhyped as a panacea to reduce greenhouse gases. The company had always been careful to say that EVs had a role in eliminating carbon emissions. But execs were always clear to point out that this rush to all electric was too much, too soon.

In 2020, Japan's government declared its intent to be carbon neutral by 2050. It wasn't an especially ambitious goal, one that many other countries have committed to. Still, Toyoda railed against the plan. At the time, he happened to be serving as the head of Japan's auto-lobbying group. He issued an ominous prediction: if the government's rules for cutting greenhouse-gas emissions took hold, 8 million of the roughly 10 million vehicles the county produces annually would go away. More than 5 million jobs would go with them, Toyoda warned.[31] "If they say internal combustion engines are the enemy," he said, "we would not be able to produce almost any vehicles."[32]

He went on: a move to all EVs would hurt Japan's economy because of infrastructure costs alone—as much as $350 billion would be needed to support all those cars plugging in, Toyoda said. He also warned that mandating electric cars would render them "a flower on a high summit"— too expensive for the average car buyer. "When politicians are out there saying, 'Let's get rid of all cars using gasoline,' do they understand this?" he asked at a news conference in Tokyo.[33]

Toyoda's rhetoric was considered fear-mongering by EV advocates, who finally and fully turned on the company.

"Toyota is stuck in the past, betting the climate crisis, public health, and economic benefits won't push consumers toward cleaner, cheaper EVs," said East Peterson-Trujillo, clean vehicles campaigner with the Public Citizen's Climate Program. At the 2022 NASCAR Cup Series Championship race in Phoenix, where Toyota was competing, the nonprofit group had a plane circle the event with a banner that read: "Want exciting? Drive electric. Want boring? Drive Toyota."[34]

Environmental critics ascribed Toyota's position to the fact that its market-leading hybrids were so good for the bottom line. At the same time, some shareholder groups also were unhappy, worried that Toyota wasn't thinking *enough* about its bottom line, as it continued losing customers to Tesla in the most important market—California—in the profitable North American market. Former Toyota owners were Tesla's largest source of fresh customers. About 8 percent of Tesla's new owners nationally were Toyota defectors, according to research firm Strategic Vision. Investors were rewarding the go-fast companies like Tesla and BYD with big stock-market valuations, not the old-guard company staking out a middle-of-the-road position.

"Toyota's approach puts it at a competitive disadvantage compared to its peers, which have shown a far greater commitment to transitioning to battery electric vehicles," said a New York City official in June 2022. The city's pension fund had about $140 million invested in Toyota.[35]

Akio Toyoda was convinced that his strategy was the right one, but the criticism of Toyota's environmental commitment bothered him. He took it personally, worried that it would sully his legacy.

"Please don't attack us in that way," he pleaded with media members at a race event in late 2021. "The enemy is carbon," he said, "not the internal combustion engine. We're not going to limit ourselves to just one option."[36]

Toyoda was so concerned about how the media played Toyota's story that he had the company create its own news outlet, the Toyota Times, to get his viewpoints out directly, without scrutiny or filtering from the media.

The postpandemic EV boom coupled with the sharp rise of Tesla and BYD sales, profits, and valuations surprised Toyota insiders at the company's gleaming glass headquarters among the lush green hillsides in the Japanese city that bears the company's name—Toyota City. Akio Toyoda couldn't ignore it. He tapped one of his most trusted lieutenants—engineer and top executive Shigeki Terashi—to examine Tesla's strategy and how Toyota should approach EVs.[37]

Then in his late sixties, Terashi was one of Toyota's most respected and longest-serving executives. He was stationed in the United States. for much of his forty-three-year career. He ran the company's large, 1,200 person R&D center in Ann Arbor, Michigan. He has been in charge of power train, supply chain, and EV strategy. After Toyoda and Elon Musk shook hands at Musk's mansion in California in the early 2010s, agreeing to codevelop an EV, Toyoda turned to Terashi to manage the project.[38]

This time around, Terashi's team dove in on Tesla, analyzing everything from its supply chain and charging network to battery strategy and model lineup. The team's conclusion was astonishing: Tesla had leaped ahead of the industry in just about every way.[39]

Tesla had reached profitability on EVs so quickly by bringing in-house almost all the components that Toyota and other automakers had outsourced, not just the batteries, but the motors, wiring harnesses, and other electric innards. Then, Tesla was able to leverage its cost base against a narrow lineup of models—more than 90 percent of its global sales came from just two vehicles, the Model 3 and Model Y, which were nearly identical mechanically. Tesla kept its cost base lower by forgoing the dizzying array of trim levels, feature combinations, and model-year changes that Toyota and other traditional carmakers long had offered. And guess what? Its customers didn't seem to miss all that.

Its charging network was a huge advantage in convincing consumers to switch to electric. And those chargers allowed Tesla to scrape reams of real-world data to help it figure out important patterns in battery life and owners' charging habits, helping it to constantly improve the product; Terashi's team found Tesla was savvy about aerodynamics, a critical aspect of wringing out better highway range. It cut factory costs with its so-called Giga Press, a machine the size of a small home that stamps large pieces of the car, such as the front or rear underbody, reducing dozens of parts to a single component. Toyota—whose manufacturing brilliance has been the subject of books and business-school case studies—was scrambling to catch up on that innovation, too.

Terashi's conclusion was deflating for a company that for decades had been viewed as (and viewed itself as) the world's most innovative and best-run automaker. The gap Toyota needed to close was huge. Toyota needed to think more like Tesla and less like Toyota in nearly every aspect of its business, Terashi concluded. On the heels of his report, the company began reworking its EV strategy, including possible plans for a clean-sheet electric platform that would reduce costs.[40]

In December 2022, around the time Terashi was putting together this sober dispatch for his boss, Akio Toyoda showed up for a media event at a racetrack in a rural area of eastern Thailand. He was in a reflective mood. As cars zoomed by on the track, Toyoda insisted that his middle view wasn't some rogue, cynical thing, but a position that actually represented the "silent majority" inside the auto industry. He reiterated: the singular focus on EVs was too narrow and too soon. But he hinted that his stamina for fighting his lonely battle might be waning.

"Who is going to continue to do this?" he asked. "Until more comrades emerge, am I going to do this until I collapse?"[41]

Toyoda then grabbed Koji Sato, an engineer who was then in his early fifties, with boyish good looks and a shock of dark hair. Sato had worked on hydrogen fuel cells and led the development of the first fully electric Lexus. Amid the roar of race car engines, Toyoda asked Sato: "Can you do me a favor? Can you be the president?" Sato laughed, somewhat confused.

A few weeks later, Toyoda announced he would hand the wheel to Sato as the company's president and CEO, while Toyoda himself would stay on as chairman. Neither man would say that Sato was appointed to steer Toyota closer to the industry's more EV-heavy consensus view. But Akio sounded a note of resignation.

"When it comes to digitalization, electrification and connectivity," Toyoda said, "I personally feel that I belong to the older generation."[42]

Ironically, it was right around then that something unexpected started to happen. Toyota's hybrid sales were heating up again. Some of the bold EV predictions from its rivals weren't panning out. Toyoda may have been right all along, even if no comrades emerged to join him.

9

The Dealer Turns Down a Deal

t was a sweltering day in June 2021. The air conditioner hummed inside an office at Claude Burns's Chevrolet-Cadillac dealership in Rock Hill, South Carolina A third-generation car dealer in his mid-sixties, Burns was dressed in his customary slacks and button-down shirt, tensely leaning forward. Lanky and handsome, with gray crowding his auburn hair around the temples, Burns was intently eavesdropping on a phone conversation between his son—also an owner of the store—and their General Motors dealer rep. It was the latest in months-long discussions with GM managers about the future of their Cadillac franchise.

By that point, GM had made clear that it was betting the company on electric cars. GM's stock was finally rallying as investors gobbled up every morsel of EV news the company, or any company really, put out. Mary Barra had made headlines a few months earlier with her pledge to eliminate nearly all internal-combustion sales by 2035. That summer she would stand on the White House lawn to cheer on President Biden as he signed an executive order calling for up to 50 percent of US vehicle sales to be EVs by decade's end.

And the luxury Cadillac brand was to be the "tip of the spear," as GM executives put it. Cadillac, they said, would aim to be *fully* electric by 2030, the fastest timetable of GM's four North American brands.[1]

Burns Chevrolet-Cadillac, a half-hour drive from Charlotte, North Carolina, was one of roughly 850 US Cadillac dealerships that were being given an opportunity to ride along on this electric journey. But it wouldn't come free. Each dealer would be required to make costly store upgrades to accommodate an influx of battery-propelled cars. Beefier, stronger car lifts that could run $40,000 apiece were needed to hoist the heavier EVs in the service bay. Charging stations would be required, which usually triggered requisite upgrades to a store's electrical system if utility crews needed to tear things up to get more power in from the street. Overall, the typical cost to prep a Cadillac store for its EV future is around $250,000. For Burns, it would be tens of thousands more than that.

For most dealers, that cost is a nuisance, not a deal breaker. The average US dealership in 2021 made about $3.4 million in pretax profit.[2] If they needed to spend a quarter-million to future-proof their store to sell EVs, most would.

But GM had put a poker chip on this table that made Burns's decision—and that of hundreds of other Cadillac dealers—an agonizing one. GM field reps had been quietly approaching dealers with a buyout offer. GM was willing to stroke checks between $500,000 and $1.5 million for dealers to relinquish their Cadillac franchises.[3]

It was a calculated bet by GM brass, who needed its dealers to fully embrace electrics. It couldn't compete with Tesla if its dealers were making a half-assed effort to sell EVs. If you took the buyout, that signaled you probably wouldn't have put in the effort needed to make EVs work. If you invested in upgrading the store, you had an incentive to go all in on EVs.

There was also an ulterior motive. Executives saw an opportunity to use GM's EV transition to fix a long-festering problem: it had too many dealerships. Cadillac's US stores outnumbered German rivals like Mercedes and Audi by around three to one, while selling many fewer cars. Hundreds of Cadillac dealerships are effectively small-town Chevrolet stores that "sell Cadillac out the back door of a Chevy store," as a former Cadillac boss once derisively put it. GM executives have grumbled privately over the years that those meager sales numbers weren't enough to justify the cost and headache of keeping those smaller dealers in the network.

For Cadillac, an EV makeover just might offer the brand's best, last hope for survival. Established in 1902, Cadillac once so dominated the luxury car business that its name still stands as a figure of speech to con-note best in class—"the Cadillac of Health Plans." But Cadillac stumbled starting in the 1970s, its reputation hurt by boring designs and subpar quality. The rise of German luxury players, such as BMW, Audi, and Toy-ota's Lexus brand, chipped away at its lead. By the late 1990s, Cadillac had become the butt of jokes among auto writers and enthusiasts.

GM has been trying earnestly to resuscitate the brand ever since—from two- hundred-mile-per-hour performance models to $110,000 Escalades popular among NBA players. Executives even moved Cadillac's headquar-ters to New York at one point as a cachet play, only for the brand to limp back to Detroit four years later. Despite some success for the Escalade, nothing has fully revitalized Cadillac's image. The brand remains off the radar for most affluent car buyers, who gravitate toward Lexus, Mercedes-Benz—and now Tesla.

Claude Burns's Cadillac dealership is physically attached to his larger Chevy store that moves more than a hundred new cars a month. The Ca-dillac business fared better than many of the rural Cadillac franchises that GM had been seeking to euthanize through its buyout, but it was no sales juggernaut. Burns had been selling ten to fifteen Cadillacs a month and wasn't making much money on those.[4] Still, his Cadillac customers were loyal, and they generated business for his service department, a profit center for any dealer. And there was still some prestige factor despite the years of decline. Even though Cadillac is a faded brand among the car conscious, it still means something in places like Rock Hill.

Burns was torn. He had dismissed GM's first offer out of hand. It was around a half-million dollars. GM reps returned again and again—sometimes to Claude, other times to his son James. Eventually, the offer reached $1 million. Factoring in about $300,000 in EV upgrades that he could sidestep if he took the buyout, Burns had a $1.3 million decision on his hands.[5]

"I think Mary Barra is making a mistake. She's going too far, too fast by going all electric," Burns said, in his deep southern drawl. "But what if she's right and I'm wrong? And the internal-combustion engine goes away?"

Burns Chevrolet-Cadillac was founded in July 1923 by Claude Burns's grandfather, Claude W. Burns. His granddaddy, as he called him, lost an eye in an accident with a mule whip as a young man, which prevented him from serving in World War I. He instead became a mechanic and later began selling cars in nearby Lancaster, South Carolina, where he opened a dealership. Burns's father would take over the location in the late 1950s. Claude began working there at age ten, picking up cigarette butts from the service-bay floor. "My dad didn't appreciate child-labor laws," Burns quipped.

Claude eventually took over, and sales remained steady for decades. Chevy's market share nationally shrank over that time, but population growth in the suburbs of Charlotte was more than enough to offset that decline. Burns doesn't have the same emotional attachment to Cadillac, a franchise he acquired around 2000. Still, he likes the status it brings to his operation.

Burns counts plenty of reasons to be skeptical about how quickly people in Rock Hill, South Carolina, will warm to electric cars. It's not San Francisco or New York—or even Dallas. People tend to drive longer distances here. There are fewer charging stations. The ideological and political divide in a place like South Carolina runs deep, he acknowledges. Plenty of his customers see EVs as a pawn in a liberal agenda and flat out wouldn't consider one regardless of whether they could afford it or had good charging options.

Burns thinks GM's EV sales targets are way too ambitious. There are factors that people haven't thought enough about that he thinks will naturally curb EV sales. What happens when someone gets into a fender bender and damages the battery, totaling the car? He thinks insurance rates will soar. And, he believes, the industry's appetite for this transition will ebb and flow based on politics and which party controls Congress or the White House, which could cause GM to backpedal.

Burns is so convinced of the staying power of the internal-combustion engine that in recent years he invested $1 million in a quick-lube facility to pump out oil changes and sling air filters, things electric cars will never need.

And yet, despite all his doubts, Burns still can't shake his hunch that the EV thing could have legs. He might doubt the timing, but he does think EV sales will gradually become a larger part of the business. And he's worried he'll get burned if he sells out just before Cadillac takes off.

Burns, an avid deep-sea fisherman, might be from small-town South Carolina, but his world view spans far beyond Rock Hill. While traveling, he has often found himself in the back seat of a Tesla Uber and always peppers the owner with questions. "To a person, they love it. They'll never go back," Burns said. He also knows that Wall Street sees growth in electric cars. And that means the automakers need to be fixated on them, too. This was all weighing on his decision about whether to sell out.

"If the stock goes up just because Mary Barra talks about electric vehicles, and Cadillac is the tip of the spear, well, then, why would I give up Cadillac?" he said. "My thinking is, I need to hedge my bets."

Burns had done a lot of this back-and-forth in his own mind. But it all came to a head that day in his office in June 2021, leaning over to hear his son's call with the GM rep. Burns fumed as his son held his iPhone up on speaker. The GM rep on the other end had changed terms. If the dealership wasn't going to comply with GM's new EV requirements, the rep said, the automaker could just strip the franchise, without any compensation at all. In reality, this might have just been a hardball negotiating tactic that GM was using as it tried to prune its network of Caddy dealers. Under franchise laws, an automaker can't yank a dealer's franchise without plenty of just cause. But this strong-arm move was enough of a trigger for Burns.

"That's when I just exploded," Burns recalls. "I said, 'Nobody ever said we aren't gonna do what we're supposed to do! We're going to do it. We're keeping the franchise. Quit calling and tell your boss not to call us again!'"

· · ·

Claude Burns's blowup is a likely precursor to the battles that will erupt in coming years between legacy automakers and their dealers, which operate nearly 17,000 dealerships across America and employ more than 1 million workers. The car-dealership business model hasn't changed much in a century, ever since franchise dealerships sprouted in the 1920s and '30s. GM, Ford, and Chrysler—the dominant US players through much of the twentieth century—didn't want to put up the capital to open their own stores all over the country. Franchising allowed them to quickly build out national dealership networks to sell cars, then service and repair them.[6]

These businesses operate separately from the brands they sell. They buy the cars from the automaker—"the factory"—and take on the risk of selling down that inventory. For an automaker like Ford or GM, their customer technically is not the car buyer, it's the dealer: the vehicle is considered sold as soon as it's shipped to a dealership. The automaker-dealer relationship is often tense, with bickering over everything from the number and type of vehicles that are sent to the stores, to the promotions carmakers use to move the metal, to what color floor tile the dealers lay in their showrooms.

Increasingly, dealerships are owned by large holding companies, some publicly traded. But the industry remains fragmented, and many are still family-owned—businesses like Burns Chevrolet-Cadillac—passed down through generations. They sponsor local youth baseball teams and robotics clubs. They also have immense political clout. Chances are that a city council candidate or county clerk got campaign contributions from their local car dealer. Many of those local officials go on to statehouses and to Congress. As Claude Burns puts it: "There's a car dealership everywhere there's a lawmaker. We support them."

It might sound straight out of a bygone era—and many people wish it was. A 2016 Harris Poll found that 87 percent of people disliked at least one aspect of the automotive retail experience, and 61 percent felt ripped off in some way.[7] Especially in the post-pandemic period, when shopping and paying for things online became much easier, more car shoppers than ever wonder why they need to step foot inside a dealership to buy a car through an awkward, often unbalanced negotiation.

The transition to electric vehicles is threatening to finally force major changes in the dealership business model. For starters, it's expected to put a big dent in dealers' service business. EVs require less maintenance than internal-combustion-engine cars, because of the relatively small number of moving parts.[8] All of Burns's oil changes would go away, along with transmission flushes and spark-plug sales. Brakes should last much longer, because many EVs offer "one-pedal" driving, which allows the driver to both accelerate and slow down using only the accelerator pedal under many conditions. A 2020 *Consumer Reports* study found that lifetime maintenance and repair costs for EVs will run about $4,600 on average—half the $9,200 expenses of a gas-engine car.[9] That's money straight out of the dealer's pocket.

The digitization of cars also allows more fixes to be done remotely, rather than through a dealership visit. Tesla pioneered the smart car, building its vehicles from scratch with a central computing brain that manages the vehicle's various functions. Traditional automakers are following that model.

Now imagine that Ford or Toyota conducts a safety recall to fix, say, a power window that runs the risk of a short that can cause a fire. Instead of sending customers to their dealership for a $400 fix—paid for by the automaker—they zap a software update to the cars themselves. Huge savings for the car company. Huge loss of revenue for the dealer.

But there's another, more existential threat to the dealer model lurking in the EV revolution. Some car executives see it as an excuse to get out from under the model all together, a model that made much more sense fifty or a hundred years ago than it does today. The legacy automakers salivate over Tesla's ability to sell directly to customers without the middleman. And they're envious. Tesla buyers go online, review the price, and click "buy," with no haggling, no two-hour wait, no checking with the manager, no hard sell for a wheel-and-tire package.

Already in Europe, more automakers have switched to a so-called agency model, where the dealer essentially becomes a sales agent. They don't buy the car from the factory, but they get a commission for every sale. This means the dealer stocks very few cars and the automaker gains much more control over final prices and customer interactions.[10] Ask a sales exec at an automaker in the United States what they think of the agency model, and most would say—not loud enough to be heard—that they'd switch in a heartbeat.

A few are vocal about usurping dealer control. Ford's Jim Farley has publicly lamented that Tesla has a $10,000-per-vehicle cost advantage over Ford—$2,000 of which he attributes to Tesla's ability to sell without a dealer.[11] One of Farley's first pushes as CEO was to come up with a new dealership model specifically for EVs with talk of no-haggle pricing and even no inventory stored at dealerships. It freaked dealers out.

"We've got to go to a non-negotiated price. We've got to go 100 percent online," Farley told an investor conference in 2022. "There's no inventory. It goes directly to the customer. And 100 percent remote pickup and delivery."[12] If you're a car dealer, he might as well have insulted your mother.

US dealers have a powerful force field in state franchise laws, which were crafted by the same legislators that those dealers have spent years cultivating relationships with. Those rules guard against traditional car companies cutting their dealers out of the process, and the strong political ties of the dealer community will help auto retailers fend off change that would hurt their businesses. But it seems clear that the arrival of electric and increasingly digital vehicles will disrupt the status quo, and many automakers and consumers are eager to see that happen.

. . .

Around the time Claude Burns was blowing up at his GM rep and vowing not to take a buyout, another dealer twenty-five miles down the road was also considering the fate of his Cadillac franchise.

In the rural old textile town of Lancaster, Gary McWhirter, Burns's first cousin, was locked in his own months-long negotiation with GM. The dealership was in the same location where Claude W. Burns—grandfather of both men—established his first store nearly century earlier. McWhirter's Cadillac business was even smaller than Burns's store; he sold a couple a month, relying almost exclusively on his Chevrolet business. His was exactly the type of small-potatoes Cadillac franchise GM was eager to snuff out.[13]

A few years older than Claude Burns, McWhirter had a similar career arc: he began sweeping floors at the Lancaster store in the 1960s, eventually graduating to washing cars and doing some cashier work. Around

1974, while attending the University of South Carolina, McWhirter's mother called him and his brother one day. The economy was tanking, and the store was having a hard time making payroll. "She said, 'You two boys need to come back to the dealership,'" McWhirter recalls.

McWhirter returned and started selling cars. He did go back to school to get his business degree, but the dealer thing stuck, and he eventually took over the store. At the time, it sold Chevy, Oldsmobile, and Cadillac. McWhirter says he watched his Cadillac clientele slowly dwindle over the decades as the cars became dull and uninspiring. He recalls the universally panned Cadillac Cimarron from the 1980s, the brand's first small car—and infamous for being little more than a warmed-over Chevy Cavalier.

When his GM rep called to float the idea of selling his Cadillac franchise, McWhirter was willing to listen. Even as Cadillac's lineup had slowly improved over the past two decades, McWhirter's client base didn't. Still, some of his most loyal customers—mainly doctors and attorneys—counted on him for their Caddies, and he worried about letting them down and losing prestige in his community.

Then he started thinking hard about what the next decade would look like as the brand transitioned to EVs. "If we go all electric, here in the South, how long is that gonna take before people gravitate to electric cars?" he recalled later. Along with the uncertain economics of the switch, there was also the certainty of the 250 grand in upfront costs. "I think it's a ten-year deal before we see our first dollar of profit," he said.

On the last day before GM's deadline, McWhirter picked up the phone. He agreed to sign away his Cadillac franchise after more than sixty years—a half-million-dollar bet that his Chevy gas-engine business would be just fine for years to come.[14]

10

Farley Channels Elon

T he conclusions from Shigeki Terashi's sobering report to Akio
Toyoda on Tesla's EV lead would not have surprised Ford CEO
Jim Farley. It was around the same time that Farley ordered the
twin autopsies on a Mustang Mach-E and its direct competitor, the Tesla
Model 3.

Despite his top engineers' confident proclamations that there was effectively no difference between the design of the two vehicles, the evidence
of Tesla's preposterous lead over Ford—and everyone—lay there on the
floor. Hundreds more parts and hundreds more meters of wire weighing
hundreds more pounds lay strewn like entrails under the Mustang. The
Tesla design was better and more efficient, cleverer in many places. The
Ford looked jury-rigged by comparison. Farley started to see how Tesla
was racking up profit margins around 20 percent on these cars, twice what
Ford made on gas-powered cars.[1] And Ford was still bleeding money on
EVs. In the end, Farley estimated the Tesla held a $3,000-to-$4,000 cost
advantage, simply by having a cleverer design for its electrical innards.

The head of Ford's engineering who Farley says assured him that the
Mach-E's innards were competitive with Tesla's was Hau Thai-Tang, a
widely respected executive inside Ford. He was credited with leading the
launch of a revered early 2000s generation of the Mustang muscle car. He
even landed on Elon Musk's radar, who early on had considered Thai-Tang

as a potential CEO for Tesla, according to Tim Higgins's 2021 book, *Power Play: Tesla, Elon Musk, and the Bet of the Century.*

Thai-Tang maintains he never misled his boss on the Mach-E-vs.-Tesla comparison. An earlier teardown of the Mach-E by an outside firm found that it *was* competitive with the Tesla on its *material* costs—the expense involved in sourcing the parts, he says. That was an altogether different exercise than the humiliating early-2022 autopsy to examine the *design* costs, a measure of how efficiently the parts fit together and how easily they could be assembled.[2] Nonetheless, Thai-Tang found himself on the bad side of a CEO who was plenty comfortable firing even popular executives. The thirty-four-year Ford veteran would retire a few months later.

Farley was so shaken by that teardown, he brought Ford's entire board of directors through the warehouse a few weeks later. "I wanted the board and my team to understand that there is only one ethos for this transformation, and that is transparency," he said. "We need to be humble. We need to understand that they are a generation ahead of us in engineering elegance."[3]

. . .

Meanwhile, Elon Musk was busy dancing.

In March 2022, Musk flew to Berlin for the opening of an assembly plant. Tesla shindigs follow the same, highly effective playbook: thousands of enthusiastic fans, many of them handpicked; hip music booming; social-media influencers hanging out; and Elon, arriving fashionably late. When he finally takes the stage, his stilted, sometimes meandering remarks in his faint South African accent never quite seem to match the event's hype. But the crowd cheers anyway, as if U2 had just come back for its encore.

This party followed the template and included some of the three-thousand-plus local workers hired to bolt together the cars. The fact that Tesla would pump out Model Ys and 3s right there on German soil was a significant milestone and surely a terrifying thought for European car executives. Six months earlier, the Model 3 had become the top-selling model in Europe—not the new top-selling *electric* model, but the most popular new vehicle, period.[4] Most of them came from Tesla's new Shanghai plant, which was built in less than a year, one of the fastest construction times ever for such a big factory.[5] It was the first time in history that a car built outside of Europe—the birthplace of the automobile—topped the continent's sales charts.[6]

The $5.5 billion German factory had the capacity to produce a half-million Model Ys and Model 3s a year, about 1,300 cars a day.[7] It was Tesla's third assembly plant—or was it the fourth? It was becoming hard to keep track. A second US factory opened around the same time, in Austin, Texas. With the German plant coming online, Tesla now would be cranking out EVs in the world's three largest car markets. And yet the company still struggled to keep pace with demand. The wait-list for a Model Y in the United States at the time stretched to eight months.

At the German fete, Musk delivered thirty of the first German-built cars to customers. He signed the vehicles and shook buyers' hands, as a bemused German chancellor Olaf Scholz looked on. As the shiny Model Ys rolled through the bright columns of white lights that workers used to inspect the cars for blemishes and imperfections, Musk danced to the beat to the techno background music and his clapping audience, bobbing and swaying, slightly offbeat, fists clenched.

On the day of his Berlin boogie, Tesla's stock rose for the sixth straight session, again eclipsing the $1 trillion market cap. Fewer than ten companies had ever hit that mark—an ultra-exclusive list that included Amazon, Apple, and Microsoft.[8]

For a long time, Wall Street had assigned such a rich value to Tesla based on its growth potential, despite heavy losses. Now, Musk was combining growth with those big margins and big profits that had Jim Farley ripping apart cars to figure out. A few weeks after the Berlin event, Tesla posted a record profit of $3.3 billion for the first quarter, in the same league as automakers that sold five times as many vehicles. The growth story remained intact too: revenue leaped 80 percent from the prior year, to nearly $19 billion.[9]

Tesla was firing on all electrons. Farley was worried. But he also believed Ford had a better chance than any legacy car company to close the gap. His secret weapon? A guy who encourages his employees to wear bunny slippers.

. . .

Even before the notorious autopsy, even before he took over as CEO, in October 2020, Farley knew Ford was behind. Frequent trips to Silicon Valley in the preceding years to chat up tech executives had convinced him the auto industry was a massive target for disruption. Every big tech company on the planet was trying to get into cars. Apple and Google had their touchscreen interfaces in millions of vehicles. Intel and Nvidia computer chips were enabling features that allowed for hands-free driving. Microsoft was sucking data from the car to the cloud. And, of course, there was the direct threat of Tesla, an actual car company.

If Ford was going to compete, it needed a serious talent upgrade. "I knew pretty quickly that I couldn't wait for our team to know what to do," he said. "We were so far behind Tesla."

Farley asked some Ford board members for contacts with recruitment people in the tech space. He laid out ten attributes he was looking for to help Ford across all aspects of an EV transition. He wanted someone who was comfortable writing software for embedded systems—essentially computers that go into mobile devices, like smartphones and vehicles. Also, someone who knew batteries and EV components, and who had intimate knowledge of battery supply chains.

The recruiters returned with a handful of partial matches, solid on some of Farley's asks. But a person who checked all ten boxes? There were almost none. But "this Doug Field guy's name keeps coming up," Farley would say later.

Few car execs would have known the name Doug Field at the time. And it's not like he was a rock star in Silicon Valley, either. But at that tightening nexus between tech and mobility—where cars are morphing into rolling smartphones—it would be hard to beat Field's bona fides.

After getting an MBA and a master's in mechanical engineering from MIT in the early 1990s, Field began his career at Ford, of all places, designing car parts. He eventually made his way to Silicon Valley, where he spent almost a decade at Segway, the two-wheeled personal transporter company that was among the most-hyped tech startups of the late 1990s. A half-decade at Apple followed that, where he was vice president of hardware engineering for the iconic Mac computer. Then he did a five-year stint at Tesla, where he led the development of its breakthrough Model 3. In 2018, Field returned to Apple, to head a secretive, long-rumored car project.

The fact he was in the upper echelons of the Valley working on some of its most iconic products meant, in short, that a career detour to Dearborn, Michigan, probably wasn't in the cards for this guy. And Farley knew it. He just wanted to start a dialogue, which wasn't all that unusual. Detroit and Silicon Valley talk a lot. Partnerships, joint ventures, licensing agreements—deals are always being done.

Farley was struck by the rarity of the combination of talent he needed. Even tech-savvy automotive people still didn't understand the type of elegant design interface that Apple routinely nails. At the same time, software engineers at big tech companies didn't appreciate the complexity of cars. Field bridged that gap, and he had chops in electric-cars and batteries. Farley asked one of the people who had helped him narrow his list to make a connection, and that person nearly laughed in his face.

"He's at Apple. He'll never call you back," Farley was told.

But Farley managed to make it happen through a mutual connection. Once he had Field on the phone, Farley made clear immediately that he wasn't prying into Field's work at Apple or his experiences at Tesla. He told Field that Ford was trying to solve lots of problems—how to how to turn cars into digital, updatable devices; how to engineer EVs without breaking the bank; the best way to develop a battery supply chain; how to design its own semiconductors.

The men chatted about all these topics and more, including the prospects for self-driving cars (they both harbored skepticism in the near-to-medium term; this conversation would influence Farley's decision to eventually dissolve Ford's autonomous-car group).

Farley could tell that Field not only had an intimate knowledge of these tech conundrums, but he had solved some of them. One of Farley's most urgent priorities was how to consolidate dozens of electronic-control boxes scattered throughout the car to operate different functions—one for brakes, another for the engine, and so on—into a centralized brain. Field had helped do that at Tesla.

"I started to understand what he was talking about," Farley would say a few years later. "He educated me."

The conversation went well enough that Farley went to California and met Field on a weekend at Moffett Field, a military air base in Mountain View, California, with nicely appointed conference rooms and meeting spaces, a twenty-minute drive from Apple headquarters. It was the first of many in-person visits. Field even attended one of Farley's vintage-car races, at Raceway Laguna Seca, near Monterey.

After several months, Farley eventually called the original mutual contact to thank him for the connection. As he was about to end the brief call, the contact said:

"Uh, Jim, you didn't ask the question."

"What question?" Farley asked.

"Well, why didn't you ask him if he wanted to work at Ford?" the person said.

"Because he would never work at Ford!" Farley responded.

"Jim, why do you think he's spending all this time with you?"

Farley was floored—and excited. But just as quickly, his excitement fizzled. "I can't afford him," Farley told the go-between. "He'll make more than *me*." This person encouraged him to at least explore the idea.

In a brief phone call with Field, which both men later described as awkward, Farley floated the question: Would he ever consider working at Ford? "Yeah, probably not," Field said initially. They kept talking. Field asked if the position would be based in Michigan. Farley said it would.

"I'd have to learn more," Field told him.

"Is that a maybe?" Farley asked.

"Yeah, it's a maybe," Field said.

Farley began sketching out a job for Field as the clunkily named chief advanced technology and embedded systems officer. He'd make Ford's future EVs updatable, able to zap new features remotely from the cloud. He'd spearhead Ford's assisted-driving technology, which takes over steering, braking, and other functions in certain situations. And he'd figure out how to piece together the insides of an electric car in the most efficient way possible.

The CEO quickly went to Ford executive chairman Bill Ford and the rest of the board to make the case for a compensation package that could lure this Silicon Valley guru. Farley knew that the numbers would be stunning for a car-industry chief technology officer. But he was adamant that talent was the top variable that would determine whether Ford could pull off this transition faster and more successfully than rivals, like GM. The board bought in right away.

Field's stock-grant-heavy contract would pay him $10 million in his first full year at Ford, and more than $15 million in his second year—making him the highest-paid executive after Farley (who made $22.8 million and $21 million in those years, respectively).[10]

. . .

Unlike his blunt, hard-charging new boss, Field was understated and quiet. Tall and lanky, he gave off a cerebral air. But he was an iconoclast from the outset. He showed little patience for drawn-out meetings and PowerPoint presentations. He favored flannel shirts and sneakers, casual even at Ford, where jeans and crisp button-downs were the unofficial uniform. "I don't care if they come to work in bunny slippers, but we've got to have the best people," Field said shortly after arriving at Ford.[11]

During one meeting early on, his team was discussing the use of a specific component in Ford's EV design when one engineer piped up: "That would be really expensive. We can't do that." According to an executive who was there, Field calmly responded, "I have to say, great engineers never talk that way. Great engineers deal with constraints, and cost is one of them." The room went silent. It was clear that the new boss expected them to solve problems.

Later, Field reinforced his lofty expectations at a town hall meeting when he veered onto the subject of rock-n-roll drummers. He explained that there have been plenty of great drummers spanning the decades. And then there's the otherworldly drummers, like the late John Bonham of Led Zeppelin and Rush's Neil Peart. These guys, Field explained to thousands of Ford workers, were obsessive about their craft. The message seemed clear: be obsessive with the task at hand. Also, being ultra-talented doesn't hurt.[12]

"With Doug, the really talented people are fine," Farley would say later. "Other people struggle."

Field brought two central viewpoints that would shape Ford's approach to this do-or-die transition to electric, digital transport. One was a conviction that, because the battery is so damn expensive, everything else around it—the wiring, the motors, and yes, the fasteners—had to be engineered with a miser's frugality. It was a minimalist ethos drilled into Field at Apple, where founder Steve Jobs famously tried to ban physical buttons from the company's iPhones and tablets. And it was a hallmark of his time at Tesla. The company was so close to death when engineering the Model 3 that cost containment was imperative.[13] Every tiny widget, every gram of content, had to earn its way into the car.

Field's other mantra: Ford needed to embrace the idea that it was selling a device that could get better over time through frequent, seamless software updates. Other automakers already had adopted this smartphone-on-wheels concept by the time Field joined Ford, in the fall of 2021. But Field brought unmatched experience to the task. Tesla had been zapping remote updates to its cars for many years. As early as 2013, Tesla beamed down a software update to raise the ride height of its Model S sedan at highway speeds, after it was found that road debris had punctured a car's battery pack.[14] It would be a decade before traditional automakers could even dream of pulling off something like that.

In the future, automobiles will be defined by experiences inside the car, the ability to change and improve over time, Field believes. Exterior design will always be important. But it didn't make sense to waste money the way the auto industry had long done, through costly tweaks like a new taillamp design or slightly modified grille to signal a new model year. Instead, Field espoused a future in which owners will get digitally delivered enhancements for years after their purchase, a radically foreign concept in the car business, where the business model has always been: make a car, sell it, and hope the customer comes back years later to buy another one.

"When you take an iPhone out of the box, it's the beginning of the journey," Field told the *Wall Street Journal* in his first interview after joining Ford in September 2021. "The traditional automotive industry has said that's the end of the journey."

Cars continue to take over more chores for drivers, and many insiders believe self-driving cars are within reach. If that happens, the movie screen or computer interface will matter more than how the car handles going into a sharp turn, Field says. He channels his Apple roots when describing this vision of a living room on wheels.

"A company that does this right could theoretically take the wheels off and plop it in your backyard and it's a product," Field said. "It becomes fundamentally the best place you could have a conference call or listen to music or watch a movie."[15]

Field's job, in essence, was to make sure the next generation of Ford electrics, coming out by 2026, would come as close to that vision as possible and catch up to Tesla in the process.

He acknowledged the steep learning curve. He referenced the difficulty that legacy companies have when trying to transition away from one product that has served as their bread and butter for generations. It's a conundrum famously laid out in the 1997 book *The Innovator's Dilemma: When New Technologies Cause Great Firms to Fail*, by Harvard professor Clayton Christensen, which posits that successful, established companies have a disincentive to forge into new markets. "There's a lot of books that say this will be very, very difficult for Ford to be a part of that future," Field said. "I'm here for that challenge."[16]

. . .

Field's first challenge came from within. His idea for a spartan, minimalist EV power train was too simple, some engineers complained. Field also questioned some of Ford's conventional manufacturing practices, including one that existed forever inside the body shop at Ford's assembly plants. When a vehicle's fenders, doors, roof, and other exterior panels were welded together, workers typically also welded on a front and back panel. That created an inconvenience for workers further down the line, in the final assembly area, because it made it harder for them to access the mechanical guts of the car. Tesla kept those front and rear panels off until later, allowing workers to essentially walk inside the car to attach their parts, saving time.

When Farley learned about this, he was exasperated. "I'm like, 'Holy crap. Why would we do that?'" The CEO was told, essentially, that it always had been done that way. Factory supervisors said they never thought there was another option. "This is, like, the craziest goddamn thing!" Farley would say later.

But inside an automaker the size of Ford, such protocols were drilled into engineers and plant managers over generations. Big changes are hard to make and are sometimes considered borderline blasphemy. There's an expression in the car business for when executives arrive from the outside and agitate to change the status quo: the antibodies start to reject them. It's often quiet or passive-aggressive resistance. Mid-level managers might drag their feet, hoping the executive in charge will leave or switch jobs. Either way, Farley could feel the antibodies mobilizing to thwart Field's mandates.

"We were having all of these shitty conversations about why we couldn't do all these things. I was like, 'We can't win this way,'" Farley would say later. Around this time, the CEO would sometimes use an analogy to drive home his concerns. "Tesla is like a 300-kilometer-per-hour Shinkansen [Japanese bullet train] that just came into Tokyo Station," he would tell people. "And I'm like, on one of those pushcarts. I'm never gonna catch him using the Ford way."

A few months after Field's arrival, heading into the holidays in 2021, Farley hatched an idea to circumvent the institutional resistance. He began calling his executives to float what he characterized as "the biggest decision I make as a CEO." He said to them, "I'm thinking about splitting up the company."

Wall Street had been urging automakers to consider spinning off their electric-car divisions to, in investor-speak, unlock value. The idea was to give investors a clean bet on Ford's EV-and-digital innovations without having to tote all of its twentieth-century industrial baggage: union contracts and pensions and gas-engine factories that would one day need to be written off and closed. Ford and other carmakers examined with their bankers what such a separation would look like. Most quickly dismissed the idea. There were too many interwoven pieces of the business. The engineer who designs door handles or seats or an infotainment system does it for both an EV and a gas-powered car. The supplier of those components doesn't care about the type of power train in the car they go into, either. Nor does the person who screws the car together—untangling union contracts to execute a spinoff would be a messy task indeed.

But there was one elephant-in-the-room reason why a spinoff didn't make sense: the old, internal-combustion-engine part of the business was making all the money to fund the move to EVs. In 2023, Ford pulled in about $14.6 billion in operating profits from selling vehicles that burned gas or diesel fuel. Its nascent EV business lost $4.7 billion.[17]

Instead, Farley moved ahead with a reorganization in a way that didn't mess with the ownership structure. He divided the company into three new, internal divisions: Model E, which would contain the EV-development team and all the digital stuff run by Field; Ford Blue, the core entity that makes internal-combustion-engine vehicles; and Ford Pro, the unit that sells tons of beefy V-8 pickup trucks—and, increasingly, some electric ones—to landscapers, contractors, and government agencies.

These distinct business units created transparency, so investors could at least see the financial state of the various businesses. It also set things up in case a spinoff made sense down the road. But a big reason was to give some autonomy for Field and his team to move fast.

Field was busy plucking talent from Tesla, Apple, and other tech firms, too. Within six months, he had poached Alan Clarke, a longtime top Tesla engineer, to run EV development. Field hired a Google executive to oversee digital design and another tech-industry veteran to head silicon design, moving Ford closer to developing its own computer chips, a key to Tesla's tech success.

"Doug Field," one Ford executive said, "is a cheat code for recruiting."

. . .

As Farley chased Musk, the two men kept up an unusual volley of public banter, veering between playful barbs and mutual admiration. In 2021, Farley on X/Twitter congratulated Musk's *Time* Person of the Year award. The next year, Musk lauded Ford for selling its 150,000th Mach-E.[18]

The men have also exchanged snide remarks. In April 2021, Farley took a veiled jab at Tesla's advanced driver-assistance features, which automates steering, speed and other functions and has sparked safety concerns among regulators. Farley compared that to a similar system Ford was introducing at the time, called BlueCruise.

"BlueCruise! We tested it in the real world, so our customers don't have to," Farley tweeted.

Musk responded, "I found some footage of the drive," and posted a clip from the movie *Tommy Boy*, in which Chris Farley, the late comedian who was Jim Farley's cousin, screams as he drives out of control on the wrong side of the highway.

Beyond Tesla's elegant engineering, Farley admired another thing about the company: its ability to keep things simple, from its nearly nonexistent marketing to its click-and-buy retail experience. But perhaps above all, he admired how Tesla had kept such a concise vehicle lineup. For a company like Ford to survive this transition, it could no longer be all things to all people, Farley believed. It would be too expensive to keep offering dozens of nameplates, each with an expansive menu of trim levels and options.

After all, Tesla's success is built on the Model Y and Model 3, accounting for more than 90 percent of its global sales volumes. And those are essentially the same car under the sheet metal, just with different body styles.[19]

"They're kicking everybody's ass," an executive from one of the Detroit automakers said, "with basically one-and-a-half models."

Tesla had gobbled up swaths of market share without even competing in major categories of the car market. In North America through 2023, Tesla had no three-row, seven-passenger SUV, which had become America's de facto family hauler. Until the late-2023 rollout of the Cybertruck, Tesla offered no pickup—another hugely profitable staple of the business. In Europe and China, Tesla got by without a small car, the lifeblood of the auto market in those regions.

The car companies have asked the same question readers are probably asking now. How? Why has Tesla had so much success with so little choice? It forced them to start questioning some time-tested truths of the car business.

Ask a Tesla owner their favorite thing about the car, and they well may mention its updatability. Or the AutoPilot system, which takes over driving in certain situations and gets new capability added over time. Or features like Sentry Mode, which flips on a 360-degree video surveillance of the car's surroundings; or Dog Mode, which sets the cabin at a pet-friendly temperature with a message on the touchscreen that reads: "My driver will be back soon. Don't worry! The AC is on." The way Doug Field sees it, the most important aspect of car ownership should be this ever-changing digital experience inside the car, rather than the contours or colors of the sheet-metal hardware in which it's wrapped.

That world view turns on its head the ways in which car brands have competed for generations. It was always about how a car drives and how it looks, and how those things are updated every year: a new front end; a bigger, slicker touchscreen; a horsepower boost, like Lem Yeung and Steve Penkevich concocted for that Ford Fusion. Every four or five years, it might get a complete top-to-bottom redesign. During the early 2000s, in the United States alone, GM offered more than sixty nameplates across nine brands and a broad spectrum of prices, still channeling legendary CEO Alfred P. Sloan's motto of "a car for every purse and purpose."[20]

Of course, EVs still need to look cool and drive well. But, because they unlock a greater ability for carmakers to digitize the experience behind the wheel, it's no longer about how the thing handles on a country road, or the styling of the front fascia. Increasingly, it's those software updates, the digital look and feel, the apps that will inform the vehicle's identity and the owner's experience. People go years between mobile phone upgrades, but the device is constantly upgraded through software.

"You don't show off your phone anymore," Field said in 2021. "My differentiation is in things like my Twitter page or my Instagram. It's a very different way that people create their identity, whereas in the past, a car was a big part of that. I think that will largely fade away."[21]

There is, of course, a purely economic reason why Farley and other traditional auto executives are thinking more narrowly about their vehicle lineups in the EV age: cost. An individual nameplate—say, the Ford Escape—can have $500 million or more in expenses attached to it, from the designers and engineers to supplier contracts and marketing budgets. Even in good times, those giant structural costs have squeezed profit margins in the auto business. How are car companies supposed to maintain all that *and* build out a completely new electric lineup, with super costly batteries? The short answer is, they can't. Selection on the dealership lot is going to shrink in coming years.

. . .

By the fall of 2022, Farley, Field, and their team had plenty to feel good about. Ford was on pace to finish the year as the number-two electric-car seller in the United States (though still outsold nearly ten to one by Tesla).[22] The Ford Lightning was getting plugged by Jay Leno and Jimmy Fallon. Farley's decision to split the company into three divisions was helping Field's team to move fast on the next generation of EVs. And concrete was being poured at Ford's massive factories to make batteries and electric pickups in Tennessee and Kentucky.

But the eye of the needle that Farley & Co. needed to thread was small. Running a car company is tough even without all of the swirling existential questions. Now he was trying to manage two worlds: the nuts and bolts of the traditional car business and an electric and digital revolution.

The trouble was that Ford was struggling with the nuts-and-bolt aspects of the business more than most. It was issuing safety recalls for its cars at a greater frequency than just about any other carmaker.

In early 2023, Farley struck a confessional tone on an earnings call with Wall Street analysts. The company missed out on $2 billion in profit the prior year because it failed to execute. Ford's costs were bloated compared to rivals. Farley had failed to fix quality problems. He noted the irony: yes, EVs were new—but these problems were happening in the core business, the sort of block-and-tackling tasks that a 120-year-old car company should have figured out.[23] How could investors have faith in Farley's plan for the future if the company couldn't manage the present?

"We are executing a double transformation," he said. "It's hard work. As with any transformation of this magnitude, certain parts are moving faster than I expected, and other parts are taking longer."[24]

The EV transformation was the part going better than expected. But even as he spoke, speed bumps were popping up for that part of the business, too. That same week, Ford said a battery fire in a Lightning stored on company property forced it to halt production. The factory would sit idle for five weeks as Ford and battery partner SK investigated the cause.[25] Customers who had been waiting patiently for their electric truck were growing restless.

More problematic, though, were early signs that the muscular pricing Ford and other automakers were enjoying on their early EVs was beginning to fade. That same week, Ford cut the sticker price on the Mach-E by an average of more than $4,000. Farley didn't want to take prices down. But he was left with little choice.[26]

Tesla had around 60 percent of US market share in EVs at the time. When a company that dominant decides to jack around its prices, rivals are basically at its mercy. And Elon Musk was on the warpath. That huge profit margin that Farley had been marveling at during the humiliating vehicle autopsy a year earlier? Tesla decided to weaponize it.

During a four-month span in early 2023, Tesla cut Model Y prices by more than 20 percent. That meant a Model Y—its top seller, which had started around $60,000—could be had in the high $40,000s.[27] The Ford Mach-E is a direct competitor. Ford would spend the first half of 2023 pushing Mach-E prices lower, spilling ever more red ink on each sale.[28]

Other rivals were also forced to slash prices on their EVs, including Hyundai and Kia.

Musk's unpredictable price war was unsettling enough, but there were more distressing signs. By late spring of 2023, some industry sources were reporting a sudden buildup of electrics on dealership lots. Some dealers were grousing that they were going to start refusing EV shipments from the factory. Just a few months earlier, that would have been unthinkable—every EV that came off the assembly line had an immediate buyer.

That summer, Ford would pour gasoline on the industry's smoldering concerns around the health of the EV market. It cut the Lightning price by a hefty $10,000, to around $50,000.[29] On the surface, the move was a head-scratcher. The Lightning wasn't pressured by Tesla's cuts, because at the time, Tesla didn't sell a pickup truck. Rivian was just about the only real competitive threat at the time, and that was a very different product. So why take down Lightning prices by so much?

Ten days later, Farley would explain the move this way during a conference call with analysts: "There are plenty of consumers" for the Lightning, he said. "The issue is the price they're willing to pay has come down."[30]

There you had it—an explicit acknowledgment that, while tech-bro types and other early adopters were willing to roll the dice on Ford's first electric truck, the company needed to appeal to more mainstream buyers. How? More range, a cooler design, better tech—but most importantly, lower prices. Farley knew those things were coming on the next-gen truck. But that wasn't due out for a few years. How would Ford be able to grind through this relentless price war in the interim?

Under the dark storm clouds of all that uncertainty, Farley in August 2023 decided to unplug from the daily CEO grind and reconnect with customers. He wanted an on-the-ground view in the epicenter of the US EV market, California. His three-day trek in a Ford Lightning was equal parts market research, photo op, and family road trip (his teenage son, Jameson, tagged along).

In Bakersfield, he visited a dealership where he used the Lightning as a generator to blow up a bouncy house for a dozen screaming kids.[31] He stopped at public charging stations to chat up EV owners and hear their war stories. He joined workers at a wind farm to see how they were using the Lightning in their operations.

He then drove south to Los Angeles, where he stopped at one of Ford's biggest dealerships, Galpin Ford, to deliver a pickup truck to Dwayne "The Rock" Johnson. Farley, dressed in a blue Ford Lightning–branded polo, khaki shorts, and gym shoes, spent about a half-hour excitedly showing the Rock the ins and outs of the electric truck. The actor nodded intently—and then drove off not in a Lightning, but in a new F-150 Raptor R, a 700-horsepower off-roading brute with a giant 5.2-liter, supercharged V-8 gas engine, thank you very much.

While in LA, Farley met up with a TikTok influencer and tried to impress him by doing doughnuts with the EV pickup in a Walmart parking lot. He visited a few shops that specialize in ripping out the gas-engine insides of vintage cars—from Broncos to Rolls-Royces—and converting them to electrics. Farley was in his element—behind the wheel, meeting real customers, talking cars.

Driving out of town, he spent over an hour chatting with a reporter sitting shotgun. Farley acknowledged that the EV story had gotten complicated in the past ninety days, with the price war and the mixed signals on consumer demand. But the thing that spooked him most happened a few months earlier, half a world away.

Farley that spring had taken his first trip as CEO to China—the country had been on Covid lockdown for years before authorities eased restrictions. He and his CFO, John Lawler, made the rounds to Ford's factories and visited managers and business partners. Then they stopped at Ford's longtime joint venture partner, Chang'an Automobile Co., a two-hour flight from Shanghai, in central China. There, the American executives were treated to yet another vehicle teardown. This time, it was Chang'an engineers dismantling a BYD vehicle. Farley was blown away. The simple, ultra-efficient design of the EV power train was frighteningly similar to what he had seen on the Tesla more than a year earlier.

"Large castings, no brackets, super simple integrated engineering, radical simplicity at the component level," Farley said. "It was all there."

And the Chang'an engineers said they had adopted the same approach on their new vehicles. This business partner, which for more than two decades had leaned on Ford for technical expertise, had surpassed Ford in EV sophistication.

Farley was staring down an existential question. While timelines could be debated, he remained convinced—along with many other auto execs and other smart people who think about these things—that the global car market was barreling toward an all-electric future. But if relative newcomers like Tesla, Rivian, BYD, and God knows how many other Chinese automakers could pull off such high-quality products, where was Ford in all of this? What would Ford be able to bring to this race that would allow it to thrive for another 120 years?

Farley got one opinion, at least, from his partners at Chang'an as he walked away from that vehicle teardown.

"They were like, 'Yeah, we don't need Ford anymore.'"

11

From Pistons to Pickaxes

A t first glance, the Salton Sea appears as a shimmering jewel in a barren, otherworldly landscape. Stretching for thirty-five miles in California's southeast corner, near the Mexican border, the young lake was formed around 1905, when the Colorado River broke through an irrigation canal and spilled into the lowest point it could find, the Salton Basin. The deluge into the basin lasted two years, resulting in California's largest lake, which looks like it could serve as a tantalizing oasis, a reprieve from the oppressive desert heat.[1] For a time that's what the Salton Sea was. Over the years, it has been an important way station for migratory birds such as oystercatchers and plovers on their seasonal journeys from Alaska to Patagonia and back. In the 1950s and '60s, Southern Californians—including celebrities like Frank Sinatra—would drive several hours to sunbathe and water-ski in its salty waters.[2]

But driving along its vast and empty shoreline today, it doesn't take long to realize something is off. Faded signs pointing to the beach, boarded-up hotels and crumbling gas stations tell a story of a once-bustling, glamorous tourist destination that time forgot. Because of how it was formed, the lake has no outflow to the ocean, so in theory it should just evaporate over time, and it does. But it hasn't disappeared as fast as it would on its own, as it is fed water, mostly by agricultural runoff, thick with pesticides, and fertilizer. All that evaporation and runoff has concentrated the body of water into a toxic stew.[3] The birds are mostly gone. Nearly all of the fish are, too. The Salton Sea's salinity levels were once similar to the ocean's 35,000 parts per million. Now they're nearly twice as high, at 60,000 ppm.[4] The receding lake has left vast stretches of salty, chemical-laced seabed that blow up into plumes of dust that locals and health experts say has worsened medical problems like asthma for the communities in the Imperial Valley, where one in five live in poverty.[5] The Salton Sea is at once beautiful, poisonous, peaceful, and eerie.

But a few new people have descended on the Imperial Valley in recent years. The scrubby, sunbaked landscape around the southern end of the Salton Sea happens to be one of the world's largest sources of geothermal brine. About a dozen geothermal power plants have sprung up in the area to tap that to produce roughly 700 megawatts of electricity, enough to power a few hundred thousand homes.[6] The boiling-hot brine, which is not connected to the large inland sea itself, gurgles below the surface of the desert floor. The liquid is pumped out of the ground and into tanks, where energy from its intense heat is converted to clean steam. That, in turn, spins large turbines, creating electricity that is sent to a generator and then offloaded.

Underground, the brine, a gray muck that reaches 500 degrees or higher, is so hot that it dissolves some of the minerals and metals from deposits around it. That means, after it wends its way through the power plant, the leftover brine is chock full of elements like sodium, manganese, silica, iron, and others.[7] For decades, operators of these plants have treated the end-state brine as waste, and much of it was pumped back into the ground. But, by the mid-2010s, the value of one mineral mixed into that slurry had shot up: lithium.

Lithium, the key metal in most of the batteries used today, is the lightest metal on the periodic chart and plentiful across the globe—the US Geological Survey has pegged the total at 98 million tons. But only about a quarter of that has been identified.[8] The problem is, it's difficult to extract. There are two main ways of removing lithium from the earth: from hard rock or salty brines.[9] Either way, the lithium is always found mixed with other elements in compounds. Getting the Li means breaking down the compounds, which can be complicated and expensive.

Roughly 60 percent of extracted lithium is mined from rock formations, with Australia, China, and Canada as leading producers.[10] Mining for the mineral is energy intensive and not environmentally friendly. It can use as much as three times the amount of carbon during the process as does extracting the material from brines.[11]

A big chunk of the global lithium stripped from brines comes from the so-called Lithium Triangle in South America, a vast area of salt flats across Argentina, Chile, and Bolivia. To separate out the lithium found in liquids, operators have traditionally used a method called solar evaporation. Salty groundwater is pumped to the surface and pooled in giant, shallow ponds spanning many square miles, and left to bake in the sun for months. Eventually it becomes a concentrated salt mixture, and the lithium is sifted out. This is neither efficient nor sustainable: it's a major drain on local groundwater reserves, and often less than half of the lithium is recovered.[12] In areas of the Andes mountains where these ponds operate, vast rows of salty metal waste piles stretch for miles.[13]

The brine near the Salton Sea was inviting innovators and entrepreneurs who might find a better way, for good reason. California governor Gavin Newsom has called the Imperial Valley "the Saudi Arabia of lithium."[14] By some estimates, there is enough lithium mixed under the desert floor there to supply 40 percent of the world's demand for electric cars—as much as 600,000 tons of lithium carbonate annually.[15] That's enough to power more than 10 million EVs. Historically, that much lithium would be valued at $7 billion to $8 billion. But by 2021, lithium prices were behaving anything but normally, with investors fixated on all things EV. Prices spiked to more than $70,000 per metric ton, which meant that those dozen or so geothermal power plants dotting the desert floor in the Imperial Valley were sitting on a $40 billion potential windfall.[16]

The price of lithium did recede some from those lofty levels, but the expected steady rise in EV demand—one research firm predicted a fivefold increase in lithium use between 2023 and 2030—has kept prices historically high. In the United States alone, the business of lithium could grow sixfold, to $55 billion annually by the end of the decade, as battery makers and car companies seek domestic supplies to qualify for billions in federal incentives in the Inflation Reduction Act if they source lithium domestically.[17]

The Salton Sea is a twenty-first-century version of the California gold rush.[18] Among its earliest pioneers was Eric Spomer. The Denver native was born into the oil-and-gas business. His dad was a petroleum engineer, and Spomer's career started as a "land man," working on minerals acquisition and lease agreements. But in the early 2000s, he switched to the renewable-energy business, not the sexy kind, like solar and wind, though. He worked on projects converting landfill gas into energy in upstate New York and a biomass operation, where scrap wood was taken from logging operations and burned to create electricity.

Spomer, who's in his early sixties and bald, with a slight frame and sharp wit, also had experience in geothermal energy. He often dreamed about the Salton Sea because of its sheer size and potential, and began scouting potential sites for a geothermal plant in the early 2000s. He mentioned his Salton Sea aspirations to an attorney with whom he had been working on a biomass project in the Northeast.

"Oh, my God, don't go to the Salton Sea," Spomer recalls the lawyer saying, unsolicited. "It'll break your heart. And your bank account."[19]

It was a harrowing place to drop a geothermal plant, desolate, scorching, and expensive. The brine pumped up from thousands of feet underground is so hot, salty, and choked with metals, so corrosive, that it basically requires building the equivalent of a wastewater treatment plant adjacent to a steam turbine to make it useful. Spomer's team had experience stripping out minerals from other hydrothermal operations, though, so they went for it, and began developing on the southern edge of the salty lake, about forty miles from the Mexican border.

Spomer spent years lining up loans to build the $400 million plant and arranging agreements with utilities for the electricity it would generate. It opened in March 2012.[20] The slurry is pumped up into towering beige tanks, six stories high, to hold the boiling brine. Plumes of steam twist up from towering emissions stacks, as the brackish liquid is sent through a maze of giant pipes that eventually pump leftover water back into the ground.

Spomer's lawyer was prescient. The economics of Spomer's geothermal power station worked for a while, because power producers in California were required to generate some of their electricity supplies from renewable sources. But while he was getting up and running, the price of solar power continued to drop, weakening the market for geothermal and other renewables. By 2015, the business case for Spomer's gamble in the sweltering desert was drying up like the Salton Sea itself. He needed to salvage the operation.

Spomer turned his attention to something he'd never tried: mineral recovery.

At first, the goal was to pull out zinc, which could be used in fertilizers, and manganese that could go in fighter jets and other military applications. But as his team was hashing over those plans, around 2015, the EV market began to stir. "That's not why we came to the Salton Sea. We had come to do power," Spomer said. "But all of a sudden all anyone wanted to talk about was lithium. And we're sitting on a lot of it."

It was settled. Spomer's company would add a second business on site: making lithium for EV batteries. There was a problem: nobody knew how to get the lithium out of that brine.

. . .

A shortage of batteries looms as one of the biggest roadblocks on the path to a societal transition to electric cars—or, more precisely, a shortage of the raw materials needed to make those batteries. It's right up there with the vexing lack of charging infrastructure, and the ability of the electric grids in developing countries to handle all those EVs plugging in.

The auto industry's rapid pivot to electrics in recent years sparked a global scramble to lock in mineral supplies a decade out. Lithium is the workhorse metal in just about all types of EV batteries and therefore gets outsized attention from industry players and investors.

But there are other critical minerals that, if in short supply, could derail the EV push. Most of the tens of billions being spent on new battery factories in the United States as of 2024 are for lithium-ion cells that also include nickel, cobalt, and manganese. Plus, graphite is needed in the battery's anode. Copper foil is needed to transmit electrical currents. Stop the flow of any of these minerals, and that's a major problem for any car company making electric drivetrains.

By the early 2020s, it had begun to dawn on car execs that there weren't warehouses full of lithium, nickel, and graphite waiting to be shipped as soon as the ink dried on a supply contract. If the carmakers wanted to ensure future supplies of these critical metals, they'd better go out and do it themselves.

"You have to secure your supply," Carlos Tavares, CEO of Jeep-maker Stellantis, said. "If not, you're out of business."[21]

But the prospect of every automaker getting all these metals out of the ground and into their battery cells in the time frame and quantities at which they need them is bleak. The go-to source for information and forecasts on battery minerals, a UK-based research firm named Benchmark Mineral Intelligence, said the industry will need to grow the supply of lithium used in EV batteries about fivefold by 2030, to something like 3 million to 5 million tons. That would require an investment of $54 billion to $116 billion in lithium projects alone. Even hitting the low end of that lithium output would require most new projects to go off without a hitch, something that is not in the DNA of the mining business. "It's almost impossible," one Benchmark analyst summed up. "Definitely a race against time."[22]

And that's just lithium. There's also a good chance that there won't be enough nickel, graphite, and other metals if automakers even come close to their plans for EV sales by the end of the decade, experts warn.

The idea that there simply wouldn't be enough of a critical component is an alien concept for traditional carmakers. Sure, they're used to fluctuating prices for materials like steel or aluminum, which can wreak havoc on their financial forecasts. When President Trump took office in 2016 and started slapping tariffs on imported steel and aluminum, for example, it knocked a billion dollars off the bottom lines of GM and Ford.[23] But no supply at all? That's rare, usually only the result of a natural disaster like a hurricane or earthquake that knocks a certain component offline. Car executives got a taste of that in the crippling computer-chip shortage in

2021 and 2022, and immediately started making supply-chain changes so they'd not go through that again.

So car companies got into the mining business. Instead of relying on their battery-cell suppliers to handle it, automakers, one by one, decided to cut out the middleman and contract directly with the source, in some cases buying stakes in mines and other extraction operations.

GM did a flurry of globe-spanning deals, including an agreement to buy Australian-mined cobalt from a Swiss-based miner and buying a new plant in Quebec to make battery material with a Korean chemical company.[24] Stellantis sank more than $150 million into a copper mine in Argentina; the company also teamed up with Volkswagen to create a new company to dig copper and nickel from a mine in Brazil.[25]

Miners surely had detected the same signs flashing green over the years for EVs to take off: the governments around the world clamping down on carbon emissions; China's massive investment in electrification; Tesla's meteoric rise. Why didn't the mining sector mobilize years earlier to start pumping out more lithium and nickel?

The reality was more complicated. The global mining business had been hammered when the last so-called super cycle—an extended period of booming commodity prices—ended in the wake of the 2008–2009 Great Recession and as China's explosive economic growth began to ease.[26] The slowdown hurt many debt-laden mining companies, which naturally shied away from investing in big projects in the following decade. Also, not all miners were convinced this early EV boom had staying power. They'd heard the auto industry's story on electrification before, only to have it not arrive.

Some miners did bet on a boom and started projects in the late 2010s. But the average development time of mines started in that decade was more than fifteen years, meaning many won't be ready until closer to 2030. They will be needed: Benchmark Minerals predicts that as many as 384 new mines for lithium, nickel, cobalt, and graphite are required to meet EV output demand by 2035.[27]

Carmakers eventually realized that not only would the mining companies need to be assured that there would be dependable customers ready to snap up all that lithium and copper once it's finally extracted, but they also would need capital to get these projects up and running.

Auto executives didn't relish getting into this business, but felt they had to. For one, the adoption of the Inflation Reduction Act, which dangles massive tax subsidies for EVs and batteries, gives them incentive to control their mineral sourcing. They knew, too, that if they put up capital and took ownership stakes, they'd secure better pricing than they'd get on the open market. It also gave greater visibility into their supply chains at a time when sourcing of these materials was coming under scrutiny. The battery industry for years had been accused of allowing child labor and other human-rights violations in cobalt mining, for example.

So the carmakers dipped toes into a business they knew almost nothing about—and one that in many ways was the polar opposite of theirs. Car executives live by rigid production schedules and count on suppliers to deliver parts to exacting specifications. In mining, cost overruns and project delays are the norm. Often, miners don't know exactly what they'll pull from the ground until operations are finally running. Automakers are used to stable, hedged pricing. Miners live through boom-and-bust cycles. But these strange bedfellows would need to get comfortable with one another quickly. A global race for battery minerals was on.

. . .

In 2018, less than a year after GM CEO Mary Barra declared her vision for an all-electric future, a veteran GM purchasing executive named Sham Kunjur was assigned to help GM secure a steady supply of EV batteries.

Kunjur, who is from India, had started his career as a supervisor at a parts factory in India that made spark plugs and fuel pumps. Before his battery assignment, Kunjur had ascended the ranks in GM's massive purchasing division, where he bought alloy wheels and led a team buying gas and diesel engines.

For his new battery gig, Kunjur spent six months in South Korea, working closely with GM's longtime battery-cell supplier, a division of Korean conglomerate LG.[28] Early talks focused on the possibility of cranking out batteries as a joint venture. The focus was on the battery cells, not the ingredients that go into them.

"I remember seeing a report from our raw-materials team at the time, saying, 'There's plenty of lithium out there. There's plenty of nickel,'" Kunjur recalls. "'We will buy them from the open market.'"[29]

Those reports didn't anticipate the percolating EV boom, led by leaders like Barra making grand proclamations, or the sudden surge in EV sales in Europe and China. Or Tesla's market valuation soaring, which created an influx of EV startups flush with cash from an early 2020s investor frenzy. Or the legislation that led to a battery factory building boom in the United States. Not long after he was advised that the open market would suffice, Kunjur started hearing more dire predictions of battery shortages from industry analysts and government agencies.

Kunjur's focus shifted upstream from the batteries themselves to the stuff they're made of. He took charge of locking in all of the automaker's battery minerals for the next decade. Kunjur discovered the urgency of his mandate during one of his first meetings with Barra to discuss how GM would source the raw materials needed to hit the CEO's ambitious targets.

Ever polite and professional, Barra could also flash impatience, Kunjur soon found out—especially when her signature initiative, EVs, was threatened by a potential supply-chain constraint: "I thought you had figured this out yesterday," Barra told her fellow engineer.

Kunjur and his team went to mining conferences to learn, to chat up miners, and scout deals. Kunjur felt out of his element, but he did find himself bumping into plenty of other auto executives in the same situation. The sense of urgency to get ahold of finite minerals was palpable.[30]

Mining veteran Todd Malan watched this incursion of auto executives with bemusement. Malan is head of climate strategy at Talon Metals, the majority owner of the Tamarack mine, a large nickel-copper-cobalt deposit in central Minnesota. Talon was a startup in the mid-2010s when it partnered with mining giant Rio Tinto to develop the Tamarack. Their first customer: Tesla, which agreed to take 75,000 metric tons of nickel for its EV batteries.[31]

Most of the world's nickel is concentrated in Indonesia, with some also in the Philippines and Russia. The United .States barely cracks the top ten in nickel production.[32] The Tamarack mine is one of the few sources of high-grade nickel in the country.[33] With people like Kunjur from GM scrambling to find US-sourced battery minerals, guys like Malan were suddenly being courted aggressively, which made him laugh.

"I don't have to buy many lunches," quipped Malan, a fifty-something University of Washington grad. Malan enjoyed watching the culture clash between auto and mining worlds.[34] "You've got the mining people with dirt under their fingernails and some great stories at the bar about exploring the Amazon," Malan said. "You can spot the autos guys because they usually have the tech-bro vests on."[35]

The car people were quickly finding out that lining up contracts for battery minerals wasn't at all like buying spark plugs or steering columns. Contracts for lithium or nickel typically stretch a decade or longer, double the length of a typical automotive contract. And, given the industry's shape-shifting strategies for battery chemistries, the automakers were making educated guesses on which metals they might need ten years from now. What if a new battery was created that didn't use nickel, and VW or Ford were stuck with a half-billion-dollars of the stuff?

Kunjur piled up some wins. GM was among the most active carmakers in securing battery materials. Three years earlier, the company hadn't even thought about the need to do this, but now it had a team of forty working full-time to scour the globe, crossing remote mountain passes and touring mining operations from Nevada to Chile and Australia.

In July 2022, Kunjur flew to Phoenix for a major lithium conference, where he met Jonathan Evans, the CEO of a young company called Lithium Americas. The company was developing the Thacker Pass in remote northern Nevada, the largest known sedimentary lithium reserve in the United States.

Kunjur didn't know it at the time, but Evans's team was juggling more than fifty interested suitors that week, from other big car companies and battery makers to hedge-fund managers. As talks progressed, Evans explained to Kunjur that the Thacker Pass site would cost more than $2 billion to develop, and that Lithium Americas would be looking for not only customers but investors too. Evans also told Kunjur he was shopping the future supply in chunks—he envisioned up to three customers, taking around 10,000 tons of lithium each.

"We want all of it,'" Kunjur told him. A few months later, they got all of it: up to 30,000 tons—the largest-ever direct investment by an automaker in battery raw materials—and the carmaker agreed to take a $650 million equity stake in Lithium Americas. The mine is expected to start producing lithium by 2026, enough to power 1 million EVs.[36]

. . .

Back at the Salton Sea, Eric Spomer was still trying to crack this problem of how to squeeze lithium from his power plant's brine. At this point, though, he found himself competing with Warren Buffett, Bill Gates, and Jeff Bezos.

As lithium's value surged, entrepreneurs began tinkering with ways to physically separate lithium without the huge ponds used in solar evaporation. The process became known as direct lithium extraction, or DLE, and was lauded as faster, cleaner, and more efficient.[37] It uses 90 percent less land than evaporation, and the water gets pumped back underground after the lithium is stripped out, minimizing the loss of groundwater.[38] The DLE process allows commercial-grade lithium to be extracted in a matter of hours instead of months, and at lower costs. Analysts at Goldman Sachs in 2023 called DLE "a potential game-changing technology" and compared it to the shale-oil revolution, where fracking of natural gas and horizontal drilling to get at previously hard-to-reach oil dramatically reduced the US dependence on foreign oil.[39]

DLE was also unproven. It hadn't really been used on a broad commercial scale. That attracted the entrepreneurs trying to solve the problem, and deep pockets backing them.

Buffett's company, Berkshire Hathaway Energy, owns ten geothermal power plants in the Imperial Valley—all of them except Spomer's. Berkshire partnered with an Oakland, California, startup, Lilac Solutions, which has received funding from a venture-capital firm backed by Gates, Bezos, and others.[40]

Spomer was David to this Goliath, scrounging federal loans and rounding up financing in a bid to become the first DLE project in the Salton Sea. He bet on a technique honed by a Dow chemist, Charles Marston, a white-bearded Michigander who had tinkered with teasing out minerals from brines for decades. Spomer's company had been testing Marston's system for years. Toward the end of 2022, he got validation: another startup trying to extract lithium from Utah's Great Salt Lake agreed to license Marston's technique.

For Spomer, a decade in the desert looked close to finally paying off. In August of 2023, inside a portable, triple-wide trailer on site at the power plant, the air conditioner hissed full blast to fend off the 115-degree heat outside. An engineer with a grizzled, salt-and-pepper beard down to his chest passed around a glass jar, a little bigger than a beer bottle, labeled "raw brine." The hazy orange liquid, he explained, is flushed through cylinders filled with resin beads. The beads gobble up lithium chloride and pass through everything else, similar to how a Brita filter traps impurities and lets through only clean water. After that, the lithium is rinsed away and later turned into a powdery solid, lithium hydroxide.[41]

Spomer had visions of producing 20,000 tons of lithium a year using the method, enough to supply batteries for more than 400,000 EVs a year.[42] Now all he had to do was convince investors to bet $1 billion on building out the needed equipment and facilities.[43] The plan was to house the new operation on a bare cement tarmac shaded by a metal overhang, steps away from the power plant.

At least one auto executive was cheering for Spomer's success: Ford CEO Jim Farley. The uncertainty in this new world of mineral extraction was forcing Ford, and all the car companies, to hedge their bets. For Ford, Spomer's experiment in the desert was one such wager. Ford had just signed a binding agreement to buy lithium from Spomer's operation once the startup began churning out the white powdery metal.[44]

12

Southern Hospitality

Two decades before Jim Farley was tooling around California in an electric pickup fretting about Tesla and the Chinese, officials from the Tennessee Valley Authority boarded a helicopter. The executives from the massive utility that serves millions of homes across a half-dozen southern states were whooshing over the lush countryside of rivers and farmland, scanning for plots that could serve as a blank canvas for a car factory.[1]

By then, the early 2000s, the US auto-factory footprint already had already stretched south, driven in large part by Japanese manufacturers. The foreign brands gained so much market share after the 1970s oil crisis that they needed to build US factories. Honda opened its first US plant near Columbus, Ohio, in 1979. Nissan opened in Smyrna, Tennessee, in the 1980s, to build Datsun pickups.

In the 1990s and early 2000s, the South landed BMW in South Carolina, Mercedes-Benz in Tuscaloosa County, and about a hundred miles from the Mercedes site, Honda in Lincoln, Alabama. Hyundai decamped to that state a few years later.

Foreign automakers were signaling continued expansion of their US factory footprints, and TVA officials knew more car factories could be coming—potential economic boons to whatever region might land one. The chopper that day passed over Haywood County in West Tennessee, which was home to not much of note. About 17,000 residents live inside the county's 526 square miles, making it one of the state's most sparsely populated areas.[2] Its scenic Hatchie River boasts more species of catfish than just about any river in the United States.[3] The area leads the state in cotton production. Haywood County is the kind of place you can find towns with names that sound like they're from a Faulkner novel, like Nutbush, Shepp, and Dancyville. Its biggest claim to fame is native daughter Tina Turner, who grew up in the county seat of Brownsville.

The torrid development and booming economy in Nashville—about a three-hour drive to the east—has bypassed sleepy areas like Haywood County. The county's median household income is about $45,000, and one in five residents live in poverty. It's been losing population for decades.[4]

The TVA officials hovering over that green expanse didn't see all that, though. They saw an uninterrupted, flat slab of land. They saw Interstate 40 slicing past the site's southeast corner, connecting the parcel all the way to California. On another corner of the massive property, they saw a major CSX Transportation railroad line. The site was protected from flooding. And it sat atop the Memphis Sand Aquifer, which could be used to generate geothermal power.[5]

It was an ideal spot to build a car factory. The TVA placed its bet.

. . .

For a century, the upper Midwest served as the automotive industry's nerve center. The Detroit's Big Three—GM, Ford, and Chrysler—concentrated their vehicle-assembly plants and engine-and-transmissions factories in Michigan, Ohio, and other Great Lakes states. Thousands of parts suppliers sprang up across the region to be close to their customers. It brought in an influx of workers as part of the Great Migration, when millions of African Americans migrated from southern states to escape Jim Crow laws and find work in the industrialized Midwest and Northeast.[6]

In 1900, for example, Detroit's African American population was around 4,000. By 1920, it topped 40,000, and a decade later reached above 120,000. Detroit's auto factories offered wages and job security that few other cities could match. In 1920, nearly 80 percent of Detroit's African American men were employed in manufacturing or mechanical jobs, compared to 36 percent in Chicago and 21 percent in New York.[7] By 1940, Ford was one of the nation's largest private employers of African Americans.[8]

Detroit emerged as an economic powerhouse rivaling Chicago. From that epicenter, the auto sector sprawled across southeast Michigan and into surrounding states. By 1950, roughly one out of every six working Americans' jobs was dependent on the auto industry.[9]

Today, the EV revolution is prompting at least a partial reversal of the Great Migration. Southern states have attracted a disproportionate share of the new, massive battery and EV factories. More than half of the $110 billion in EV-related investments disclosed by auto companies from 2018 to early 2023 were earmarked for southern states.[10] The geographic stretch from Michigan down through Kentucky, Tennessee, and into Georgia took on the name "the Battery Belt."

"Fifty, 60 years ago, we would leave the South, going to Detroit, going to Indianapolis, going to Flint, Mich., going to Saginaw, Mich. To do what? To work for the automobile industry," Thomas Burrell, president of the Black Farmers and Agriculturalists Association said. "But now the automobile industry . . . is coming to us."[11]

The South's advantages have compounded amid the industry's battery-building boom. There are ample expanses of flat farmland, critical for battery plants, which need stable soil that's not prone to shifting topography. Battery factories also use lots of water, which the South has in abundance relative to the scarcity troubles facing the Plains and US West. Southern states boast cheap energy, too—an essential for electricity-hungry battery plants. In spring 2024, the average cost of a kilowatt-hour for industrial users in Tennessee was around 6.20 cents, compared with about 8.20 in Michigan.[12]

Significantly, the South is removed from the United Auto Workers' stronghold in the Upper Great Lakes. Asian automakers planting stakes in the South not only avoided competing with the Detroit companies for workers; they enjoyed a cost advantage by avoiding union wages. It's an edge that lasts to this day. GM, Ford, and Stellantis paid UAW workers in the low- to mid-$60s per hour in wages and benefits on average in 2023, compared with mid-$50s for Asian competitors and less than $50 for Tesla.[13]

But the Battery Belt also was born from desperation. The farming and textile manufacturing that had once nurtured places like Morgan County, Georgia, an hour east of Atlanta, was vanishing. In its place, Rivian has long-range plans to sink $5 billion in Morgan County for a massive, 7,500-employee factory. Hyundai also is betting on Georgia, with a $7.6 billion infusion for an EV assembly plant and battery factory codeveloped with partner LG Energy Solution outside Savannah to support 8,500 future workers.

These wins reflect some serious perseverance on the parts of local officials in fading southern towns grasping for a lifeline amid decades of decline. They set the table for the new automotive players, taking lonely farmland and connecting it to all that cheap power and water, stitching together utility lines and miles of wastewater pipes to service the vast sites. They practically hung a shingle: "Your EV Factory Could Be Here."

In many cases, mayors, county administrators, and state officials clung to those massive parcels for many years, hoping that a mega-employer would come along, a regional savior. In the case of Haywood County, it took two decades of angst, self-doubt, mind-numbing public hearings, political wrangling, community skepticism, and endurance.

They built it. And for a long time, no one came.

. . .

The TVA worked with local officials to designate that Haywood County plot a so-called megasite, to be set aside as the future home for a massive manufacturing complex. Of course, there were a bunch of hoops to jump through before some *Fortune* 100 corporation might come along and agree to transform the sleepy county with jobs and economic prosperity. First off, a few dozen landowners would need to agree to sell their properties. The task of convincing residents fell to Franklin Smith, the folksy, longtime mayor of Haywood County. Smith, with his shock of white hair, neatly trimmed goatee, and southern drawl, was beloved in the county, having spent more than twenty years as mayor over two terms. The megasite was his albatross, and then his legacy.

All the way back in the early 2000s, Mayor Smith hosted a dinner, set up by the chamber of commerce at a local church. Dozens of property owners showed up to hear about the grand plan hatched by the TVA and local officials. Smith wanted to gauge how many would be willing to sell. "My family won't take less than $10,000 per acre!" one man shouted, throwing out a number that was roughly three times the going rate for farmland in that area. "That pretty much set the price," Smith would say years later.[14]

Securing the options to buy those parcels would take years. Smith leaned on his friend county attorney Michael Banks for the legwork. Banks grew up in the county seat of Brownsville and remembers Smith coaching him and his brother in youth baseball. Square-jawed and handsome, Banks served in the Tennessee Army National Guard—he would one day be promoted to colonel—and had just come back from deployment in the wars in Iraq and Afghanistan when he took the Haywood County attorney job.

Banks's first task was to hit the rutted county roads to sell the megasite idea. He logged hours sitting at the kitchen tables of skeptical farmers who were holding out, trying to get them to sign options that said they would agree to sell whenever the state came calling. Many figured that was a long way off, and they could continue farming the land in the meantime. The supersized values per acre that Banks was dangling didn't hurt either.

Some farmers dug in, though. A pair of gun-toting brothers who owned several acres on the proposed site had been railing against the project. For months, they refused to grant an option on their land. One day Banks got a call. The brothers wanted to meet the mayor and Banks at 6 a.m. the next day. They set the meeting spot at the Quick Stop in tiny Stanton—population 415, and a fifteen-minute drive down the road from Brownsville.

The mayor and attorney were perplexed and leery of this predawn rendezvous. But they resolved to go. The pair pulled into the gas station a few minutes before six. The parking lot was empty. Within a few minutes, one of the brothers appeared from around the corner of the gas station, beckoning Smith and Banks to drive around. Smith slowly pulled his car up next to a massive heavy-duty Ford pickup truck. A rear door swung open. Smith and Banks glanced at each other and climbed up inside.

No pleasantries were exchanged. One of the brothers declared that the family was ready to sell an option to buy their property. Stunned, Banks began rifling through his files for the proper form. "Well heck, that's great. All I need you to do is sign right here."

"I ain't signing shit, boy," one of the brothers said. Banks tried to plead his case, glancing over nervously at Smith next to him on the back seat. The brothers weren't budging. "I ain't signing nothing. But you can go ahead and sign that check."

After a few tense moments, the mayor finally said, "Just write him the check, Michael." Banks scribbled a check for $80,000. The mayor and his attorney hopped out of the truck.

Back in Smith's car, Banks turned to the mayor: "How in the hell am I gonna explain to the state that I just signed a check for eighty grand to someone who wouldn't sign an option?"

Smith was quiet for a minute and then grinned. "I'm sure you'll think of something good," he said. Years later, the brothers kept their word and gave up their acreage.[15]

Once land was under option, the plan was for the state to bring its deep pockets to the table and acquire the property outright. At that point, Mayor Smith and other local officials could shop it to giant companies in the automotive or aerospace sectors, in hopes of landing an employer that could lift the standard of living for generations of West Tennessee residents.

In 2009, a vote was scheduled in the Tennessee legislature to authorize the state's purchase of the Haywood County land.[16] Days before, Smith heard rumblings that political support for the megasite was waning. Republicans in the statehouse were criticizing it as a Democrat-led boondoggle.

Banks and Smith sat in mayor's office inside the county building in Brownsville, dejected. The votes weren't there. Even the local state senator who represented Haywood County wasn't planning to back it. Just then, Banks sat up with a jolt. He remembered that Bill Haslam, then the mayor of Knoxville and a leading candidate in the upcoming race for governor, had touted the megasite on the campaign trail. Neither man knew Haslam. Smith grabbed the phone anyway and got him on the line. "The funding for this thing is gonna get killed. You said it was your number one priority," Smith told Haslam.

"Franklin, give me an hour and I'll call you back," the future governor told him.

Smith hung up. Banks looked at him. "That son of a bitch ain't gonna call us back! We're just a bunch of local yokels down here!" Banks said.

But Haslam called back in forty-five minutes. He had lined up enough votes from lawmakers on the east side of the state to save the project. The deal went through for the state to spend $40 million in taxpayer money to buy the sleepy chunk of Haywood County cotton farms.

It bought Mayor Smith and his trusted attorney another twelve years of angst and heartache.

. . .

The minutiae of permitting and red tape that attached itself to a property the size of the Haywood County parcel was mind-numbing. In 2013, state officials declared that the site needed a wastewater plan to make it more marketable to prospective buyers. Banks again hit the road, this time to negotiate easements along a fifteen-mile stretch for a future pipe that would route wastewater from the site to Brownsville for treatment. But the pipe was to discharge treated wastewater into the Hatchie River. Environmentalists howled, worried that would foul up the scenic waterway and all those catfish.[17]

So they moved to Plan B in 2017: the wastewater would be treated on site and pumped though a *thirty-five-mile* line stretching west all the way to the Mississippi River.[18] Banks again hit the road to secure easements—this time from 240 individual properties that the future line would pass through.

"Imagine me calling you up and saying, 'Hey, how would you like to have a big shitter line cutting through your field?'" Banks would say later. "I had to perfect that spiel 240 times."

Smith and Banks would attend hundreds of megasite meetings over the years, all while discontent in the community grew. Locals griped that the resources pouring into the fallow property could be better spent on schools or rec centers. Every now and again, a tantalizingly large company would come through and nibble at the site: a semiconductor company; Electrolux, the washer-and-dryer manufacturer; a tire manufacturer. None bit.

"People felt like we were the boy who cried wolf," Banks would say later. "They were just tired of hearing about the megasite."[19]

Patience was wearing thin among lawmakers at the statehouse in Nashville, too. Around 2015, then-Governor Haslam's administration hatched a plan to split the megasite into at least five separate industrial parcels. The idea was to make the site easier to market and was heralded by Haslam's economic development chief, Randy Boyd, a wealthy entrepreneur-turned-politician and owner of the Tennessee Smokies minor league baseball team.[20] The dream of landing a single, transformational employer was about to die. Smith and Banks worked the phones, but it wasn't looking good.

Soon Boyd himself—who came up with the plan to split up the megasite as part of Haslam's cabinet—jumped into the upcoming gubernatorial race, when Haslam would be term-limited. On the campaign trail, Boyd had to acknowledge that, as economic development chief, he had struck out on landing a buyer for Haywood County. But he vowed to fill out the site over time with multiple manufacturers and thousands of new jobs.

As his campaign was gearing up, Boyd, an avid sports fan who had run dozens of marathons, planned a weeks-long, 537-mile run from the east side of the state to the west as a campaign fundraiser. On a crisp late-October morning in 2017, he invited Banks, also an avid runner, to run a twenty-mile portion of the trek through Haywood County.

The men met in a parking lot on the county's east side, laced up their shoes, and set off on a ninety-minute run—eighty-nine of which were spent arguing about the megasite. Boyd insisted the best path for the project was to parcel it off. Banks knew how close the site planners had come to landing the big one. Breaking up the property at this point would squander a decade of painstaking work.

"We fought like cats and dogs," Banks recalled years later. The next day, the two met to finish a second, longer stretch of the run. They agreed the subject of the megasite was off-limits this time.

Four more years passed. In September 2021, word began trickling out from state officials about a big announcement on a West Tennessee project. Smith and Banks were confidentially clued in: the megasite had a buyer. A massive one. All the toil and hustle and aggravation seemed to have finally paid off.

The weekend before the announcement, Banks arrived at the University of Tennessee's football stadium in Knoxville, for a meeting of a state civic group. The host was Boyd, who had lost his gubernatorial bid but landed on his feet, serving as UT's president. Banks walked into dinner and greeted Boyd, grinning. "Did you hear the news?" he asked. Boyd laughed. "Truly amazing," he said.

It was Ford that bought the parcel as its home for the centerpiece of one of the largest projects in the history of the auto industry. Ford chose the 3,600-acre site—that's four Central Parks—to plop its first-ever dedicated factory for electric vehicles. Not just any EVs, but pickup trucks, Ford's lifeblood.

It will be Ford's first brand-new US assembly plant in more than fifty years, slated to open in 2025. It's designed to one day churn out a half-million battery-powered trucks a year, which would make it the company's largest US factory, period. The white metal structure springing up from the rust-colored Tennessee dirt seems to stretch for miles—and practically does. And right next to that behemoth will be another one, a battery plant roughly the same size. There is enough land here left over after that to build *another* assembly plant and battery factory. Ford gave the site an audacious name to match its grand scale: BlueOval City.

The project harkens back to Ford's touchstone factory, the Rouge complex in the company's hometown of Dearborn. There, in the 1920s, company founder Henry Ford built a massive city unto itself on the banks of the murky Rouge River. That complex housed a soup-to-nuts automotive operation that allowed Ford to dominate the early days of the car business: everything from iron ore and steel mills to a glass factory was located on the sprawling campus, seen at the time as an industrial marvel unparalleled in the world.

BlueOval City is nearly *twice* as big as the original Rouge campus. And a parallel project is humming along about 290 miles to the northeast, in Kentucky, where Ford plans two more nearly identical battery factories with its joint-venture partner, Korea's SK On. The battery plants will churn out enough battery cells to make 2 million EVs annually. Ford sold just under that number of total vehicles in the United States in 2022.[21]

A seven-minute drive north of the sprawling site brings you to Stanton, the site of Mayor Smith and Michael Banks's tense land negotiation years earlier at the Quick Stop. Across the street, Lesa Tard's tiny restaurant, Suga's Diner, is slinging three times the amount of fried catfish, coconut-rum chicken wings, and sweet tea than it was before the Ford project sprang from the ground. From late morning through evening, construction crews in dusty work boots clomp into the low-slung, red-and-yellow storefront, adorned with a Coca-Cola sign so faded it's almost unrecognizable. Those construction workers will soon be replaced by factory workers once the plants go live.

A Memphis native, Tard was already planning a restaurant expansion— three years ahead of the plant's opening—to accommodate the influx of lunch-break workers. She said some locals are worried about too much growth too fast, but most people welcome the infusion of money and jobs into the sleepy area. "They've been waiting for a long time for something to show up, and it's finally here," she said.[22]

Fifteen minutes further up the road from Suga's is Brownsville, which became a Ford town almost overnight. Months after the announcement, a two-story mural was painted on the side of a brick building in the picturesque downtown showing a vintage, 1950s-era Ford pickup truck below a giant Blue Oval Ford badge and the greeting: "Brownsville Welcomes Blue Oval City." Storefronts were dotted with small white-and-blue Ford signs that read: "Welcome Y'all!" At Livingston's Soda Shop, where a Blue Oval milkshake sits atop the dessert menu, patrons are greeted with a display case stuffed with Ford memorabilia, including a Mustang hub cap and toy models.

Mayor Franklin Smith would pass away six months after workers broke ground on the project he worked so long and hard to land. Smith's obituary called him "the cornerstone of the Megasite."[23] For Banks, who would deliver the mayor's eulogy, his friend's legacy looms large.

"These West Tennessee communities were becoming ghost towns," Banks said. "It's literally the saving grace of this region."

13

Battery USA

A Ford lifer named Eric Grubb found himself in charge of the building of BlueOval City. One day in the fall of 2022, he climbed up a three-story-high pile of dirt and looked out across a sea of construction cranes and dump trucks dotting what weeks earlier had been a flat stretch of western Tennessee farmland. He was hit with an unexpected jolt of nostalgia.

Grubb was in his mid-fifties, tall and burly, with close-cropped graying hair. He grew up in rural Indiana, where he was rarely indoors. He remembers building tree houses and using branches to dam a creek near his home. But he was also a math whiz, and one of his teachers suggested he consider getting an engineering degree.

One week after high school graduation, he began working at a diesel-engine factory in Indianapolis under a General Motors co-op program: he would work as an engineer in training for twelve weeks at the factory, and then spend twelve weeks at school in Flint, Michigan. After that, he landed a job at Ford. His first big project was to design and install test equipment for a transmission used in F-150 pickups trucks and Crown Victoria sedans. Promotions over three decades led Grubb to oversee Ford's construction activities all over the world, from Spain to India to China, where he built three assembly plants and an engine factory.

Even for Grubb, BlueOval was an overwhelming assignment. In roughly nine years before he arrived in Tennessee, he had overseen the construction

of 16 million square feet of Ford space across the globe. The BlueOval projects total 17.5 million square feet. His timeline: two years.

"Jesus, man, I can't believe the magnitude of what we're doing here.
And that I'm controlling it," Grubb thought, perched atop the dirt pile,
as a few thousand construction workers scurried about the dusty site on a
bright fall morning.

He thought back to that first big assignment installing transmission-
test equipment for F-150s and Crown Vics at a suburban Detroit factory in
the late eighties. Once, his dad, a pilot, visited Grubb at the factory while
he was in Detroit before a flight. Grubb's dad marveled at the sophisticated
equipment his son was overseeing and his $9 million budget. "If he was
around now to see this, he would be absolutely shocked," Grubb said. "It's
a lot more than $9 million."[1]

Try $11.4 billion. That's how much Ford and partner battery partner SK
On are pouring into the four factories Grubb oversees—Ford is kicking in
about $7 billion of it. It's the largest investment in Ford's history, and, ac-
cording to the company, the biggest single US investment by an automaker
ever.[2] It's not overstating matters to say that Ford is betting its future on this
former patch of cotton and soybean fields an hour northeast of Memphis.

BlueOval City is a representative snapshot of an entire industry racing
to catch up to two seemingly unstoppable forces: Tesla and China.

Beginning around 2017, most of the biggest car companies on the planet
came to the view—gradually, then suddenly—that the future was electric.
Next came an era of bold proclamations, with carmakers trying to outdo
one another. Hyundai said EVs would account for roughly one-third of
its sales by 2030.[3] VW said it was targeting half of its global sales to be
full electrics by then.[4] General Motors shocked people in 2021, when the
company best known for its brawny, fuel-thirsty pickups and sports cars
said it would essentially phase out internal-combustion vehicles by 2035.[5]

The ambitious timelines spurred a collective realization that everyone
was going to need lithium-ion batteries, all at the same time. There would
be fierce competition to secure them and meet the boasted goals. By the
early 2020s, Asia accounted for more than 80 percent of the world supply
of EV battery cells.[6] Car execs in Detroit and Germany didn't want to be

beholden to companies halfway around the world for that vital piece of the supply chain.

Plus, auto executives could feel their barriers to entry melting away. For a century, the complexity and cost of the internal combustion engine prevented upstarts from encroaching on their turf. Now it was becoming clear that if the heart of the car was going away—not just the engine block itself but all the attendant parts like cylinders, camshafts, and pistons, all connecting to the equally complex transmissions—the carmakers needed to own what it was being replaced with. That meant getting into the battery business, starting with constructing factories for building something other than cars, for the first time in a long time.

After decades of strongly resisting the creation of new factory space— it can cost $2 billion to get a plant up and running—the EV push has spawned a massive building renaissance. As of 2022, the world's top thirty-seven automakers had earmarked $1.2 trillion for new factories and related EV spending through the end of the 2020s, the Reuters news agency estimated.[7] That sum—roughly the size of Saudi Arabia's annual economic output—includes battery factories, new and converted assembly plants, and contracts for raw materials like lithium and nickel.[8] The Big Three Detroit car companies had built only a few brand-new factories over three decades.[9] By mid-2023, they had at least a dozen in the works.

The unfolding EV building boom points to a potential broader reawakening of US manufacturing generally. By spring of 2023, spending on US factory projects hit an annualized rate of $189 billion, double the rate from a year earlier. Building factories accounted for almost 10 percent of all construction activity—the most since the federal government began tracking the data in the early 1990s.[10]

Several factors contribute to the bounce back. President Trump's trade war began prodding US companies to think harder about bringing some manufacturing home. It didn't hurt that it's no longer as cheap as it once was to make things in China, as the country's wages have grown along with its middle class and economic prowess. Geopolitical relations with China have worsened in recent years under President Xi Jinping's more-hawkish approach to US trade.

But for the auto industry in particular, the lingering trauma of major supply disruptions in the wake of the Covid-19 pandemic served as an extra push to forge a deeper domestic manufacturing base.

The supply-chain dysfunction was rooted in the pandemic lockdowns that hit the United States in early 2020. Auto plants closed and carmakers hunkered down, slashing factory schedules. But within months, a sharp rebound in demand for new vehicles surprised auto executives. By the end of 2020, factories were humming again, and consumers were shelling out record sums for new wheels. But demand started to outpace supply because of a shortage of the computer chips to control virtually everything in a car, from the engine to the climate control to the power seats to the airbags.

Semiconductor makers had been scrambling to keep up with a wave of pandemic demand for laptops, gaming consoles, and other consumer electronics that were purchased by people stuck at home wanting to entertain themselves. To meet the demand, the semiconductor makers robbed Peter to pay Paul by slashing production of chips for the auto industry. By the time the surprise boom in demand for cars took hold, it was too late. A severe shortage of chips forced automakers to sporadically shut factories—in some cases for months at a time—as they tried to salvage output of their most profitable vehicles. Over two years, the chip crunch wiped out more than 15 million vehicles from automakers' production plans globally.[11]

The ripple effects caused the weirdest car market in memory. Dealership lots—normally gleaming with fresh sheet metal—were mostly bare asphalt. US car buyers accustomed to driving their new car away from the lot on the same day were forced to wait months for a new car, if they could get it at all. It wasn't uncommon for buyers to fly across several states to pick up new wheels, often paying thousands above the sticker price. The average price paid for a new vehicle in the United States rose by about 30% from 2019 to the end of 2022, to the high $40,000s.[12]

The upshot for automakers and dealers was their most profitable stretch in history. But it was a harrowing time for car executives, and they emerged from it with a rattled faith in the stability of global supply chains. When the shortage hit, many automakers didn't even know the names of their chip suppliers. That's the nature of the industry's skein of parts suppliers: automakers contract directly with one supplier—often a large global player—which then goes upstream to source its raw materials and smaller components. Those semiconductors likely were made in Taiwan, then sent to another Asian country, like Malaysia, where they were embedded into circuit boards, eventually winding up in a larger electronic-control unit made by another big supplier, and then shipped to the carmaker's assembly factory.

Auto executives were horrified to realize that they had such poor visibility into such a crucial supply thread. Many began drawing up plans to design their own computer chips and forge closer ties to the semiconductor companies. They also saw a parallel in batteries. If the EV-production forecasts being tossed around by their competitors were anywhere close to accurate, a severe shortage was likely to arrive later in the decade. Carmakers didn't want to leave that to the whims of a shaky global supply chain, one that goes right through the heart of China.

"Our strategy is to control our own destiny," GM CEO Mary Barra said in 2022 while discussing the company's battery plans.[13]

Most of the companies had some in-house battery expertise, with teams of chemists that dabbled with metals and mixes, and engineers who could expertly connect a big battery to the rest of the vehicles' systems, and assembly teams that could bundle battery cells into big packs for placement into the car. But actually producing batteries? The car companies would need help.

Manufacturing battery cells is a notoriously complex and finicky process. Massive rolls of foil are coated with a slurry that contains the material for the anode (carbon) and cathode (lithium-metal oxide). The foil gets unrolled and cut like brittle dough into long, flat sheets. The material is then moved into a massive oven and baked at high temperatures for hours, to evaporate excess slurry. Later in the process, the foil strips are sent to clean rooms, where separator material is woven in between layers of cathode and anode like a club sandwich.[14] Humidity levels are kept at a minuscule 1 percent, and workers wear the equivalent of hazmat suits to prevent contamination. The battery cells then go into holding racks for weeks of curing.[15] Very little of the process resembles what goes on in an engine factory or vehicle-assembly plant.

So the car companies turned to their battery suppliers for help building and operating plants. In 2018, there were only a few US factories making batteries for EVs. Within five years, more than two dozen were in operation or under development.[16] Almost every one of these massive factories were being built under the same arrangement: the car company goes in fifty-fifty with an Asian battery company—Korean manufactures like LG Energy, Samsung, and SK On, and Panasonic in Japan.

Once again, automakers everywhere except China were years behind Tesla on batteries. Elon Musk first outlined plans for its so-called Gigafactory, a JV with Panasonic built in the Nevada desert, nearly a decade before this, in 2014. Traditional automakers watched from afar with a mix of skepticism and curiosity, wondering why any carmaker would want to bring that complex, high-maintenance process in-house. But, by being so far ahead of the legacy automakers, Musk had his pick of partners and locked into cheap contracts for battery raw materials, which would eventually skyrocket in price.

GM was among the first legacy automakers to leap into the battery game. Steve Kiefer was the Detroit auto giant's longtime purchasing chief, overseeing an organization of thousands of employees that collectively bought some $90 billion in raw materials and car parts annually.

Square-jawed, with a thick head of silver hair and an affable smile, Kiefer had spent much of his career as an engineer tinkering in internal-combustion-engine power trains. In 2019, five years after Musk had made his move with Panasonic, GM was fixated on batteries. Kiefer would meet weekly with engineers and R&D executives in a "war room" in GM's engineering center in Warren, north of Detroit, where they hammered out a strategy to secure enough batteries at lower prices than the traditional arm's-length supply contracts had delivered.

Kiefer had worked closely with Korea's LG, which supplied batteries for the Chevy Volt and the Bolt EV. But, with the massive number of batteries that GM would need in the coming years, he knew a different approach was needed. GM needed to put up capital—a ton of it.

In his five years running GM purchasing, Kiefer had never taken a single supply deal to Mary Barra's senior leadership team for its blessing. But a deal of this magnitude demanded it. "I don't normally bring these things to you guys, but this is sort of make or break," Kiefer recalls telling his peers during a meeting of Barra's direct reports. "We're talking a multibillion-dollar contract for battery cells that will go into millions of EVs."

GM's executive team agreed the company needed to get in the battery business. Kiefer flew to Korea for a dinner with LG's leadership, where he floated a joint venture to build and run multiple US battery-cell factories. The Koreans were game and relieved to hear that GM wanted to put up billions in capital for the effort.

"I think the battery suppliers were a bit freaked out," Kiefer would say later. "They saw all these automakers coming out and declaring plans for 10 percent or 20 percent EV sales. They weren't sure how they'd ever be able to bring all that capacity online."[17]

. . .

The battery-factory binge would require automakers to use muscles they hadn't worked in a several decades, and some they hadn't ever used. Battery facilities would come with a learning curve.

Back at Ford's BlueOval City, the battery plant is similar to the nearby truck-assembly factory in size—and that's about it. One noticeable difference: a jungle of thick electrical cables snaking through the unfinished ceilings on each level. A battery factory sucks an ungodly amount of electricity, five times more than Ford's largest vehicle-assembly plant.[18] To manage that, Ford built its own power plant on site. Across the street, the Tennessee Valley Authority utility company built another 500 kilovolt substation, enough to power about 40,000 homes, to handle the added load flowing into the Ford property.

Those two substations would connect to smaller ones inside the factory to distribute all the power coming into the building. The smaller ones inside were the size of a tidy log cabin, like a home generator on steroids. To zap all that voltage around the building, workers strung up those black cables, as thick as the diameter of a softball. The cable comes wound on wooden spools taller than Grubb, who's six-foot-three.

In a busy year, Grubb might order around fifteen of these log-cabin substations to be used inside Ford's factories in North America. For Blue-Oval, he ordered 240 of them on a single day, about 80 for each of three battery plants. The purchasing manager back in Dearborn thought there had been a mistake.

Workers also were installing dozens of air-handling units that would pull the humidity way down, similar in size to the substations. From each unit, massive aluminum ductwork—large enough for a person to stand inside—sprawls out and across the vast recesses of the ceiling.

At any given time, there are five thousand workers on the Haywood County site. To make sure he's got manpower arriving in the waves that he needs them, Grubb huddles frequently with bosses of some of the biggest unions in the country to pull electricians, carpenters, and other skilled tradespeople from all over the country. In the vast dirt parking lot for construction workers, pickup trucks and rugged SUVs wore license plates from all over: New Mexico, Colorado, Texas, Florida, Virginia, Connecticut.[19]

Grubb knew that a steady flow of workers was critical because companies were competing on skilled trade talent as much as they would be on EVs and batteries. By then, more than $25 billion worth of battery plants similar in scope to Ford's were springing up from dusty tracts of farmland in places like Kentucky, Ohio, Georgia, Michigan, and Indiana. Billions more were being spent on new or retrofitted factories to make electric vehicles, like Ford's truck plant.

In the back of his mind, Grubb was thinking about a GM battery factory three hours to the east, near Nashville. Grubb got word that GM's project was siphoning some of his electricians, so he had to boost incentives for his workers to prevent poaching. He was also mindful of plans for a massive Intel semiconductor plant that was set to get underway nearly six hundred miles northeast, near Columbus, Ohio, which would put more pressure on the skilled workforce.

Grubb strolls through the cavernous construction site in a black construction hat, yellow vest, jeans, and work boots. He politely lets workers
slide past him on stairwells. "I don't want to impede progress," he says. But
he's also keenly scanning for any whiff that things are running behind.
At one point, Grubb walked through an upstairs section of the battery
factory that he thought already had been roofed and sealed, only to spot
some puddles on the floor. "Get somebody's ass up there and get it fixed!"
he told a manager.

Grubb occasionally will address the work force in an all-hands pep talk.
He always makes sure to remind them of one thing. "This," he says, "is the
job you're gonna tell your grandkids about."

. . .

The auto industry already was fully mobilizing on its EV ambitions. The
political winds in Washington, DC, were about to give those efforts a huge
supporting lift.

The 2020 presidential election that ousted President Trump swept in
a trifecta for the Democrats. A climate-minded President Biden took the
White House with a comfortable lead in the House and a fifty-fifty Senate.
The Dems saw a window to throw lots more money at the climate crisis.
Car executives would be shocked to learn just how much more.

The idea of a clean-energy building boom didn't get much traction a
decade earlier, when Biden was vice president under Barack Obama. But
a confluence of factors created fertile conditions for Biden to go big on a
climate package, wrapped in the promise of domestic manufacturing jobs.
For one, Trump's election four years earlier exposed deep anxieties among
working-class Americans about the defection of factory jobs overseas and
the hollowing out of once-proud small towns. Appeasing those voters had
taken on greater import. Projects like battery and EV factories—backed
with federal dollars—could help close that gap, lead to wins for labor, and
revitalize some faded sections of the industrial Midwest and South.

Meanwhile, Trump already had conditioned those voters and others
to feel like America was hopelessly behind China in manufacturing. And

there was growing anxiety among some Democrats around China's dominance in batteries, semiconductors, and other tech realms. While the two sides couldn't agree on whether Washington should be throwing money at the problem, the China threat created momentum to push through some weighty industrial policy.

For the car-guy president, EVs were high on his list. On a muggy summer day in 2021, he sat at a wooden desk on the front lawn of the White House, flanked by Ford's Jim Farley and GM's Mary Barra, to sign an executive order declaring his goal of 50 percent of new cars sold in the year 2030 to emit zero emissions.[20] Three months later, the president got an early win with the bipartisan passage of a $1 trillion infrastructure-spending package. Much funding went to upgrade crumbling roads and bridges. But the package also included some EV-friendly items, including the $7.5 billion to light up a half-million public EV chargers, mostly along highways, "so that the great American road trip can be electrified," a White House press release said.[21]

But the president and congressional Democrats had much bigger climate ambitions. They had been pushing a gargantuan, $3 trillion economic package, with a kitchen-sink agenda dubbed Build Back Better. It included the top item on the auto industry's wish list: expanding the availability of a $7,500 tax credit for people who bought EVs. It had been in place since the early 2000s as a carrot for car buyers to go green. The Biden package proposed tacking on an extra $4,500 for a total of $12,500, a subsidy that would make even China's leaders blush.[22]

But there was a catch: only EVs built in America—with union labor—would fully qualify for the extra $4,500. It was a blatant bid for Big Labor votes. The United Auto Workers union loved it. So did the Detroit carmakers, the only three car companies with unionized US workforces. Just about everyone else hated the idea.

The evenly split Senate meant that Biden's climate package would take the backing of all fifty Democrats. It was clear early on that one lawmaker would emerge as the most pivotal figure in Washington as this $3 trillion battle played out: moderate West Virginia Democratic senator Joe Manchin. During his twelve years in Senate, the former governor of the

Mountain State had seen the shift to cleaner energy sources like wind and solar power decimate the coal industry. Manchin was resolute that any climate package created US manufacturing jobs and would hopefully start to claw back China's two decades of gains.

For months, starting in fall of 2021, Manchin pushed Biden and his Democratic peers to whittle down the package. He already was uncomfortable with the vast sums Biden and Democrats were spending on Covid-19 relief efforts and other stimulus that he worried was stoking inflation. Democrats gradually pared down the measure to appease him. Eventually it was whittled to $1.75 trillion. In the end, it wasn't enough. A week before Christmas, the Democrat chose Republican sounding board *Fox News* to drop the guillotine on the president's economic-and-climate plan. "I can't vote for it," he told an anchor.

But Manchin left a tiny crack of the door open, telling fellow Dems that he would be open to a skinnier agenda that included some climate provisions, along with early child education, and health-care reforms. By spring, talks had narrowed to just Manchin's staff and New York senator Chuck Schumer's people, along with some Senate committee staffers. Manchin's only interest in using federal money on climate-related projects was to create jobs; he didn't care about the car industry's interest in EVs.[23] In fact, as one staffer who has been in many energy-policy meetings with Manchin over the years put it: "He hates EVs."

Manchin also was adamant that no taxpayer money should go toward anything that would benefit China. Given China's eight-hundred-pound-gorilla role in the global EV supply chain, that put the car industry in a tough spot. "His mentality basically was 'EVs equal China. So I'm not going to support EVs at all unless you can create policies that make sure no help is going to China,'" said another staffer involved in the behind-the-scenes discussions.

Manchin insisted that if any part inside an EV came from China—even one little widget—that would disqualify the vehicle from eligibility for the $7,500 consumer tax credit. Aides explained that essentially no car sold in the United States met that threshold. Bolts, rivets, springs—so many parts in a modern car are sourced from China, either fully formed, or in raw form to be finished later. Eventually, Manchin was persuaded to narrow the focus to the batteries. If the car business couldn't get its batteries from China, that would prod automakers to build even more factories in the United States.

But he didn't stop there. Manchin knew China also dominated the supply chain for all the minerals that go into batteries, like lithium, nickel, cobalt, and graphite. For a car to qualify for the full subsidy, it couldn't have any individual battery cells from China, Manchin's team insisted. It couldn't have a single molecule of a mineral that was extracted in China. It couldn't have any trace of a mineral that was *processed* in China, even if they were pulled from the earth elsewhere.

For months, Manchin's hard line on China nearly derailed the entire bill. His position was so strident that staffers worried it would essentially erase the $7,500 consumer tax credit. Dangling incentives for consumers to buy EVs is a core tactic to broaden adoption, one that helped give China its massive lead. EV supporters were worried they would end up in a worse position than when the climate talks started.

Schumer's staff scrambled for a Plan B. They would have to reorient the discussion away from EVs and toward Manchin's insistence that tax money would be going toward building US factories and employing workers. They pulled in an idea that had been percolating elsewhere in the Senate. The federal government already had a program that offered subsidies for manufacturing stuff that helped mitigate climate change, such as solar panels and electricity-storage boxes. What if they did the same thing for making batteries on US soil?

A handful of other Democratic Senators—Michigan's Debbie Stabenow, Colorado's Michael Bennet, and Nevada's Catherine Cortez Masto—already had been kicking around the idea of making billions of dollars available to help companies make batteries in the United States.[24] It was unusual: US industrial policy usually offers tax subsidies for construction of actual buildings, not for making specific products. The idea was that, by defraying the cost of the battery—by far the most expensive part of an electric car—automakers would be able to sell EVs at lower prices than they would otherwise. So, maybe the buyer doesn't get the full 7,500 bucks off because of the strict mineral rules, but carmakers could offer them for less.

Manchin bought in. Staffers were excited, but worried that those battery production subsidies would balloon the cost of the whole climate package. Eventually congressional bean counters put a number on the proposal: all of those tax breaks for battery output would cost taxpayers around $30 billion.[25] In context, that was huge. Credits for all of the other climate-production measures—like solar panels—were just a fraction of that amount. But some staffers who worked on the bill suspected that $30 billion was way off—to the low side.[26]

In July 2022, the massive piece of legislation with a curious-sounding name—the Inflation Reduction Act—was unveiled. The official price tag for all of the EV and energy portions—including non-automotive funding—was pegged at around $370 billion.[27]

Auto executives and their lobbyists rushed to dissect the bill. There was some good stuff in there for EVs—the $7,500 tax credit survived and in some ways was expanded. The sales cap that had phased out the credit for EVs made by Tesla, GM, and Ford was lifted. But there were lots of new strings attached.

To qualify, EVs had to be built in North America. That's not a problem for the US automakers but was seen as a thumb in the eye to foreign automakers like Hyundai, Kia, Volkswagen and BMW, which had been getting traction with their imported EVs. Other rules posed problem for just about every automaker. EVs would need to have US-built batteries to qualify for half the credit, or $3,750. Given all the battery factories in the works, that one eventually would be manageable. Then, the auto people read the minerals piece: another $3,750 would require all the raw materials in those batteries to come from North America or trade-friendly countries like Chile or Australia. While those rules would take effect in stages, it threatened to render nearly all EVs at least partially ineligible.

Automotive lobbyists at first howled. But then they began to digest the part about the battery tax credits. One lobbyist said he thought it was a typo. It offered $45 per kilowatt hour for US-made batteries. Some back-of-the-envelope math arrived at this: a carmaker that produces an EV's batteries on US soil in some cases could get more than $3,000 *per car* in battery tax credits.[28]

It soon sank in that this piece of legislation that seemingly materialized from thin air—hammered out essentially between two senators—wasn't just a consumer tax break for EVs. It was arguably the most extensive shot at a comprehensive industrial policy in recent U.S. history, one with huge implications for how the auto industry's EV business would unfold. There was no cap on how much the government might spend. That $30 billion estimate from federal bean counters would prove a lowball estimate. Analysts would peg the eventual federal spending through the law's sunset, in 2032, at more like $150 billion.[29]

GM would later tell Wall Street analysts that the tax credits from making all those batteries in the United States would be the deciding factor on whether the company could turn a respectable profit on EVs in the early years (8 percent profit margins with the subsidy; just 2 percent to 3 percent without).[30] Jim Farley said Ford stood to receive $7 billion in tax credits within the first few years for all those batteries it would be pumping out of BlueOval City.[31] One analyst said that if Tesla maxed out those battery credits—and it should—it could cut the cost of the Model Y by nearly 40 percent.[32]

Often big pieces of policy that spring from Washington take years to take root. This bill created an instant impact. Within weeks of its passage, Tesla pressed pause on a planned battery-making operation in Germany and began sizing up US sites.[33] Ford had been considering a factory in Mexico to make batteries using CATL technology, but quickly pivoted and announced a $3.5 billion factory in Michigan.

· · ·

Carmakers and battery suppliers are making these giant investments with the very real risk that new battery tech could emerge that would require major revamps of these new plants and supply chains—or even a complete redo.

There are two main types of EV batteries widely used today. One uses a cathode that combines nickel, manganese, and cobalt, sometimes called NMC; the other, which uses a mix of lithium, iron, and phosphate, is called LFP—the same type of battery that Bob Galyen used when he worked for CATL in China. Each chemistry has advantages and trade-offs.

NMC batteries are more expensive and are the predominant chemistry for EVs sold in the United States and Europe. They have higher-energy capacities, which generally means better acceleration and longer driving ranges.

LFP batteries have a lower energy density, which is why—at least early on—automakers didn't think they were good enough for American and European car buyers. But they have some significant advantages. The battery cells are roughly 20 percent to 30 percent cheaper than NMC, a big reason they caught on in China.[34] BYD and CATL are the world's dominant LFP battery producers. LFP batteries also are far less susceptible to catching fire than NMC batteries and have a longer life span.

Those advantages—especially the lower cost—have led more automakers to introduce LFP-based batteries in Europe and the United States in recent years, including Tesla, Rivian, and Ford.

For decades, battery makers, startups, universities, and government labs have been pursuing battery-chemistry breakthroughs that could dramatically lower cost, extend driving ranges, and quicken charging times. There has been no shortage of silver-bullet claims of game-changing innovations. But battery experts temper these predictions, saying it's unlikely that one of those would make a difference in this decade. A few bear mentioning, though.

So-called solid-state batteries are considered something of a holy grail in the battery world. They work much like a lithium-ion battery but use a glass or ceramic electrolyte to ping-pong back and forth between the anode and cathode, rather than a liquid electrolyte. That means much greater energy density—and potentially huge range improvements—along with reduced fire risk. A Toyota exec in early 2024 said the company "in a couple years" would offer solid-state-battery EVs with a 750-mile driving range and recharge times of ten minutes.[35] Honda, VW, and others have big plans for the tech, too.

But solid-state batteries are expensive and tricky to mass produce. Ask a group of battery experts over a few drinks at a battery-tech conference and skepticism abounds over how long it will be before cost-effective, wide-scale production of solid-state batteries.

Bob Galyen, who emerged from his CATL run as one of the industry's busiest battery consultants, believes solid state is about a decade behind on cost, safety, and other critical aspects. "It's going to be just like lithium-ion was in terms of the length of time it will take to hit the market," he said in 2024. "And lithium-ion took a long time."[36]

There are plenty of other promising chemistry tweaks that also have a long way to go. One uses sulfur for the cathode instead of nickel and cobalt, but it's prone to corrosion. Another swaps out the lithium electrolyte for sodium, obviously a much cheaper and more abundant mineral. Two Chinese startups in 2024 were even planning launches of sodium-ion EVs. But the energy density remains a deal breaker for most; it can cut driving range in half.

. . .

Elsewhere in Michigan, Mujeeb Ijaz was scrolling through the details of the Inflation Reduction Act after it landed. His eyes lit up.

A mechanical engineer and son of Pakistani immigrants, Ijaz spent his entire career in the US battery industry. As an eager twenty-something engineer at Ford in the early 1990s, he worked on sodium-based batteries. Ford was excited about the program and sold a few dozen EV trucks using the technology to a few companies in California. The program was abruptly halted after reports of a few battery fires. Ijaz was dispatched to California to go over some of the charred remains. He designed another battery pack that that would eventually go into an all-electric Ranger pickup truck. Like GM's EV1, though, the Ranger would flop, and Ijaz would spend another two decades on false starts and near misses on battery pursuits.

He left Ford in 2008 to join a Michigan battery startup called A123. The firm was betting on LFP batteries, leveraging those advantages over the NMC variety: lower cost, more durable, and safer.

Ijaz enjoyed success at A123—and had fun. His team developed some of the first LFP batteries produced for cars, including those used to power a hybrid Porsche 919 that won the famous 24 Hours of Le Mans car race and for an EV developed by Ohio State University that set a land-speed record of 307 miles per hour at the Bonneville Salt Flats in northwestern Utah.

But A123 struggled to lower its costs and was losing money. Despite receiving $250 million in federal aid from the Obama administration, it went bankrupt in 2012.[37] A year later, it was purchased by a Chinese company. Dejected, Ijaz jumped to Apple.

Ijaz had been fixated on the subject of range anxiety—stemming back to a road-trip trauma he endured on July 4th weekend in 2015. He loaded up his Tesla Model S, joined by his wife and three adult kids, and set off for Lake Tahoe, a 240-mile drive. About three hours into the trip, Ijaz started to sweat his driving range. The car showed he'd be cutting it close, but he knew the last part of the drive is up a mountain pass, which would suck down more battery than flat highways. He was down to twenty miles of range. He kept scanning the car's infotainment system for nearby chargers. Nothing. And then it happened. His car ran out of juice, just twenty miles from his destination, near the tiny mountain town of Soda Springs. He pulled over on a mountain pass in the dense Tahoe National Forest. He eventually had to have the Model S towed to get it charged in Tahoe, and Ubered his family to their destination.

The time in Lake Tahoe was great. But on the return trip, determined not to make the same mistake, Ijaz stopped at a charging station. There was a two-hour wait just to get a charger and another hour for the car's battery to replenish. His family was grumpy, and Ijaz was a bit humiliated. He had spent his entire career on electrification of the automobile. And now his family hated the very idea of it, spewing sarcasm at him. "They were like, 'Yeah Dad, this electric-car stuff is phenomenal. Love it,'" he would recall later.

For Ijaz, it was a formative experience. The next year, the pandemic hit. Ijaz was still working at Apple, but now from home. And his mind was restless and preoccupied. He couldn't get that Tahoe experience out of his head. He would spend time online researching range anxiety and EV range calculations. At one point he came across a report that outlined survey results suggesting people need at least four hundred miles of range to be comfortable with having an EV as their only car. But Ijaz knew that most people never would get those four hundred miles—variables like going up mountain passes cold weather or simply driving at 80 miles per hour on the highway will drain the battery more quickly.

"I started to realize that the range of an electric vehicle will never be successful until you can take a basic regional trip, like San Francisco to LA or Detroit to Chicago," he said.

Ijaz obsessed over this for months until one morning he sat up with a bolt. The range needs to be six hundred miles, he thought. And he was going to start a battery company to get to that.

His idea for a six-hundred-mile range car would use a novel approach: a cheaper LFP battery would deliver the driving range that most people need to commute or to get around town. To avoid the sort of Tahoe grief that Ijaz experienced, though, his company, named Our Next Energy, would stick a long-range, nickel-based battery under the hood to deliver another 450 miles. People don't normally need all that much range. The LFP battery is more than enough for commuting, running errands, and shuttling kids around. But when owners do want to leave town, they don't want to stress about it.

Ijaz moved back to Michigan and set up shop in a nondescript office park, about a fifteen-minute drive from where RJ Scaringe had stealthily met with Jeff Bezos in a Rivian conference room. In less than three years, the company had raised more than $300 million.[38] More than three hundred people work in his bustling office with a tech-startup vibe. The parking lot is so jammed most days that employees park on the lawn.

Three months after the Inflation Reduction Act was passed, and Ijaz's eyes had lit up, he would be joined by Michigan's governor as earth movers began clawing into the ground to prepare a $1.6 billion battery site for his company about thirty miles away from his Detroit-area office.

"Someone asked me, 'How are you going to compete with China?'" Ijaz said. "The way I see it, the US has always been the leader in innovation—the only thing we ever missed was doing it at scale. If we decide that now we're going to invest in scale, we already have the innovation to reestablish our leadership."[39]

14

Making Cars Is Hard

Back in Normal, Illinois, in Rivian's airy, zen-like "customer-engagement" room, the company's forty-something head of manufacturing operations, Tim Fallon, sat under bright klieg lights. Dressed in gray jeans and a navy blue collared shirt with sleeves rolled up, bespectacled with close-cropped hair, Fallon was prepping for an on-camera media interview to discuss how the Tesla of trucks was going to scale its business. How would it take that squared-off, taut truck with the unmistakable oval headlamps and the optional camp stove, and start cranking them out by the tens of thousands?

It was a sensitive subject. With 2022 winding down, Rivian already had cut in half its production target for the year, to 25,000 vehicles. The start-up's trouble getting trucks out the door had spooked investors.[1] Rivian's stock market valuation had cratered by about 80 percent from a year earlier, in the ebullient days following its IPO, celebrated in this same room. Things were tenser now.

It wasn't a problem with the truck. Rivian's R1T pickup and R1S SUV were racking up critical praise, and the company still had a backlog of 100,000 preorders.[2] The problem was growing pains, trying to transform a startup with an eye-catching truck design and hip brand to a manufacturer that could tackle the nuts and bolts of car manufacturing at scale.

Typically, the process of getting a clean-sheet vehicle from the designer's sketchpad to start of production takes more than four years. The design team might go through several clay-model versions, agonizing over tiny deviations in the sheet metal. This process can be even more painstaking for EVs because aerodynamics are so crucial—the easier the car slips through the air, the longer it can travel on a single charge. The designers work closely with engineers to make sure their clay ideas can transfer to digital models and then to industrial reality. All along, marketing people will weigh in on everything from exterior styling to where the cup holders should go.

While that's happening, you need to begin the catchy-sounding discipline of manufacturing-engineering. That is the process by which all the machinery and tooling the factory will need to build cars is planned and installed. Also, you need purchasing: one vehicle program might have hundreds of buyers, people marshaling suppliers years in advance to deliver thousands of parts on time and on budget. GM's purchasing budgets are bigger than Boeing's annual revenue.[3]

Even big automakers that have been engineering cars and running assembly plants for a century sometimes screw up a new-model rollout. Rivian had existed for more than a decade, but only as a brand, a concept. CEO RJ Scaringe and his company were new to manufacturing. They were starting out with a new factory, a new workforce, and three new vehicles, including an Amazon delivery van. There was nothing normal about what Rivian was trying to pull off at its factory in Normal.

When visitors arrive at Rivian's plant, it feels more Patagonia store than automotive assembly plant. The first thing they notice is the brightness relative to the dim feel of the typical car factory. Millions of LED lights illuminate the place like a high-end showroom. The setting reveals spotless, diamond-polished concrete floors and gleaming white walls and structural beams. Solar panels atop the roof generate some of the plant's electricity.

Interspersed throughout the factory floor are cozy lounge areas filled with greenery for workers to take breaks. The cafeteria feels more like a Silicon Valley tech firm than a Midwest car factory. Leafy vines from hundreds of plants hang down from the bright white ceiling. Picnic tables of blond wood match chairs with outdoorsy phrases like "Ski," "Kayak," and "Run" etched into the backrests. The menu features items from local restaurants and food companies. Options include tiramisu overnight oats, sweet-potato hash, and a cilantro-lime fajita tofu salad. There's a build-your-own plant-based grill. Kombucha and shots of a restorative juice made from bone broth are on the drinks menu.

Much of the plant itself feels modern and high tech. Six giant presses—the only real leftover equipment from the Mitsubishi days—stamp out large door panels and hoods. In the body shop, dozens of high-tech robots, painted bright blue, clutch big, stamped pieces of sheet metal for the pickups and weld together panels, sending sparks flying. A separate body area has much larger robots that look like mini T. Rexes, which handle much larger panels for the Amazon vans. From there, vehicles move to a cutting-edge paint shop, where more robots coat pickups and SUVs with colors so rich and unusual—with names like Red Canyon, Limestone, and El Cap Granite—that automotive magazines have taken to ranking their favorite Rivian hues.

But a closer look at the Normal plant in November 2022, more than a year after the factory opened, revealed signs of dysfunction. The assembly lines themselves move much more slowly than in a typical plant. At times the line might mysteriously stop altogether for long stretches, with workers milling about or checking their phones. On the separate assembly line for the huge Amazon delivery vans, the vehicles actually meandered along sideways, rather than nose to nose, because the area is so pinched for space. It's a feat that would make even GM or Ford queasy.

At the end of the line, dozens of trucks and SUVs were parked scatter-shot across the floor for inspections. Workers tinkered under the hoods of some. Others were up on hoists for mechanical tweaks. One SUV's tailgate was swung open, with a cross-legged worker inside tapping at a piece of interior trim with a mallet. These sort of final quality inspections happen in every plant. But there was a whole lot of it happening in Normal more than a year after the factory's opening.[4]

Scaringe needed help. Already there were murmurs among investors about whether the young CEO—having pulled off the successful IPO of this hip electric-car company—should step aside. Often the entrepreneur who created the vision, courted investors, and built the brand isn't the best person to lead the next phase, a phenomenon known as the founder's dilemma. The skill set moves from inspirational and visionary to practical, strategic, and regimented. Colleagues described Scaringe as brilliant, resourceful, and driven. But he also is prone to micromanaging.[5] That worked when Rivian had ten employees or a hundred. At this point, its head count was fourteen thousand.

The young CEO was scrambling to bring in experienced manufacturing hands. Among them was Tim Fallon, who was back in the customer-engagement room, settling in for his on-camera interview. Fallon spent most of his career doing the things that Rivian needed help with most. He worked more than a decade in the body shop at Nissan's massive Smyrna, Tennessee, factory, which makes the Nissan Rogue and the Leaf EV.

Fallon was prepared for tough questions. His smile was at times easy and natural, chatting before the lights went on in his Tennessee drawl. At other times, he seemed a bit nervous, his smile forced. That same day, at Rivian's Southern California headquarters, Scaringe was preparing to release Rivian's third-quarter results. Wall Street analysts already were on high alert for Scaringe to say that Rivian would fall short of its chopped-in-half production target for 2022. In trading that day, before the numbers came out, Rivian's stock sank another 10 percent.[6]

"We all feel the pressure," Fallon said once the camera was rolling. "Every day."[7]

. . .

Rivian wasn't alone. A slew of EV startups that raised billions in the fevered climate of the previous few years had seen their stock prices tank as they strained to spool up their factories. One reason Rivian's shares plunged that day was because another EV darling, Lucid, had reported the night before a much-worse-than-feared quarterly loss of $670 million.[8] That meant the company lost about $480,000 on each vehicle it delivered that quarter.

Lucid was right up there with Rivian as an EV startup with the cachet, financial backing, and polished executive team to get the attention of investors in search of the next Tesla. Lucid CEO Peter Rawlinson had been the chief engineer of Tesla's first hit product, the Model S, but left for Lucid in 2013. Like Rivian, Lucid was backed by billions of dollars from the Saudis, who were spreading their bets in their bid to diversify away from oil. Lucid eventually came out with a sensuous electric sedan, the Air. Its electric driving range was even more eye-catching, capable of going more than five hundred miles on a single charge. Just one year before that stinker of a quarter, the combination of beauty and technical prowess had vaulted Lucid's stock valuation to $91 billion, more than either Ford or GM at the time, even though it had made only a few thousand cars, volumes those manufacturers eclipse before the sun rises on any given morning.[9]

But since that high valuation, investors had started losing faith. In August of 2022, Lucid, like Rivian, halved its vehicle production forecast for the year, to around seven thousand cars. Problems ran the gamut, from supply chain to logistics. At one point Lucid had to idle its factory in the scrubby desert landscape south of Phoenix because of quality troubles on that svelte sedan. It wasn't the complexities of the battery tripping things up but the basics: Rawlinson said Lucid was having trouble with gaps between the sheet-metal panels on the hood and fenders, for example. He even dialed into the quarterly conference call from the factory floor to reassure investors that he was on the case.

Also like Rivian, Lucid's problem wasn't consumer demand. Rawlinson had assured Wall Street analysts that the company still had thirty-four thousand reservations, enough to bring in more than $3 billion in revenue and keep the plant humming for years.[10] But that would require one thing: more money.

"No one should be under any illusion," Rawlinson said. "We are under a trajectory that is capital intensive." Rawlinson said he was looking to raise $1.5 billion in equity to fund future operations. That's not what Lucid's existing investors wanted to hear, because those additional shares would dilute their holdings. The stock sank 13 percent that day.[11]

Rivian and Lucid were the highest-profile and considered the furthest along among EV startups. But there were many more that had latched onto the investor frenzy in 2020 and 2021. Not only were these upstart carmakers unprofitable, but most hadn't even started building cars or booking revenue. Normally, that would put them a long way from becoming publicly traded companies. But a financial instrument that had hardly existed a few years earlier was about to shower them with money.

The wonkily named special-purpose-acquisition company, or SPAC, came to prominence amid the brief, loose-money era starting in 2020, which included cryptocurrencies and meme stocks. SPACs essentially became a lower-bar way to take a company public. Rather than go through a rigorous IPO process that requires company executives to lift the curtain on the good and bad of the business during grueling road shows with banks and prospective investors, SPACs require less disclosure.

They work like this: a group of business partners—often executives with ties to a certain industry—forms a shell company, with just a few people and virtually no assets. This so-called blank-check company makes the rounds with prospective investors, selling their bona fides as sage industry insiders who are qualified to go find a promising startup to bet on. Investors trust these blank-check guys to pick a winner that they can merge with and take public. If the firm doesn't find a merger partner within two years, it's required to repay the investors.

It made sense that EVs were a focal point of the SPAC boom, when Tesla's valuation in 2021 crossed $1 trillion for the first time, feeding Wall Street's frenzy for EV bets. For the EV startups, a SPAC deal was a shortcut to some serious capital. The lax SPAC rules allowed companies to disclose their revenue and profit projections, a step that generally doesn't happen during the more-rigorous IPO process. The companies looking to SPAC often trotted out PowerPoints with deeply rosy projections—go figure.

For context: it took Google eight years to hit $10 billion in revenue, the fastest US startup to achieve that mark. It took Facebook, Amazon.com, and Tesla at least a decade. In 2021, though, several EV startups were assuring prospective investors that they could get there in just a few years, the *Wall Street Journal* reported.[12]

One of those was Fisker Inc. The Manhattan Beach, California, startup was founded by charming Danish car designer Henrik Fisker, who had sketched vehicles for BMW and Aston Martin in the early 2000s. An earlier startup that Fisker formed in 2007 had once been viewed as a Tesla rival. The $100,000 plug-in hybrid that bore Fisker's name counted Leonardo DiCaprio and former Vice President Al Gore as owners. But the company ran into a cash crunch after a disastrous manufacturing launch and folded in 2013.

Fisker remained amazed at the looks he would draw when he drove his aging plug-in car around L.A. Meanwhile, Tesla was finding success. Fisker couldn't resist the urge to try again. In 2016 he started another company with his wife, Geeta Gupta-Fisker, as chief financial officer. Fisker vowed to learn from the past. He would stick to what he knew best—car design and tech features—and contract out the grind of manufacturing to a big, well-known supplier Magna Steyr which also builds cars for Jaguar and Mercedes-Benz.

Fisker raised about $1 billion in a SPAC deal to fund operations. The stock's value zipped to around $8 billion in 2021. Before any cars were made, Fisker and his wife used their newfound wealth to buy a $21.8 million home in the Hollywood Hills, with sweeping views of downtown L.A., retractable glass walls, and a sauna with pink Himalayan salt walls.[13] The company turned heads with its stylish midsize Ocean SUV, priced starting in the low-$40,000s and promising snappy acceleration.

But enthusiasm for the Ocean dried up rapidly. Early reviews panned it for a slow touchscreen and other problems with the in-car tech. Worse, Fisker's back-office functions didn't seem up to the task of a complex vehicle launch. Logistical problems and understaffing delayed shipments of the Ocean from Europe to the U.S.—at one point Fisker resorted to flying over a few dozen cars, an obscenely costly move. The startup ended up selling only about 6,500 vehicles before going bankrupt in June 2024.[14]

Faraday Future was another EV startup with an abundance of optimism. Founded by a Chinese billionaire and once tagged as the "future Tesla," the California-based luxury EV maker had formed way back in 2014—and went bankrupt in 2019. Two years later, it resurfaced in the

SPAC-fueled EV craze. Faraday told prospective investors that it was projected to eclipse $10 billion in revenue in less than three years. It went public in the summer of 2021 via a SPAC merger and raised about $1 billion. By mid-2024, though, the company was warning of potentially running out of cash and had only made a handful of cars.[15]

Eventually, a few investors started calling bullshit on some of the SPAC companies.

Ohio's Lordstown Motors turned out to be one of the flimsiest EV start-ups of the SPAC era. The company was birthed opportunistically in 2019, soon after General Motors was scrambling to find a buyer for its Lordstown, Ohio, factory, which it had decided to shut down.[16] The move infuriated then-president Donald Trump, who had once assured a rally in nearby Youngstown that there would be an influx of factory jobs during his administration: "Don't move. Don't sell your house!" he said.[17]

Trump harangued CEO Mary Barra for weeks to find a buyer to save factory jobs in the politically pivotal state. Desperate to divert Trump, GM entertained several whacky proposals for the massive factory, including one plan to use it as a giant greenhouse and a bid from a local car dealer who urged GM to continue making small cars there so he could purchase them for a new Uber-like ride-hail business.[18]

Eventually, a little-known entrepreneur named Steve Burns stepped forward to buy the factory to build electric pickups. Burns had toiled at a few lesser known EV startups. GM's corporate-development people—the ones in charge of acquiring companies and new technologies—were horrified at the prospect of doing business with such an unknown entity. But GM execs wanted desperately to get the president off their backs.[19] The automaker eventually kicked in about $75 million for a minority stake in Lordstown Motors.[20]

Trump was quick to trumpet the news that GM would sell "their beautiful Lordstown Plant" to Burns's company. "With all the car companies coming back, and much more, THE USA IS BOOMING!" he tweeted.[21]

Thus began a colorful three-year trek into bankruptcy. Things started out well. Lordstown raised about a half-billion dollars in a SPAC merger in August 2020. A month later, it got national attention when Trump hosted Burns outside the White House. Trump gushed over the hastily constructed Lordstown Endurance pickup truck. "It's an incredible piece of science and technology," he told Burns before a crush of cameras.[22] By January 2021, the company said it had about 100,000 nonbinding Endurance reservations from business customers.[23] Lordstown's value soared to about $5 billion in February of 2021. That would turn out to be its peak.

The next month, a short seller—a company that invests so that it profits when a company's stock declines—issued a detailed report claiming Lordstown had duped investors. The firm, Hindenburg Research, called the EV maker's order book "a mirage," and said it paid a consulting firm to generate truck orders ahead of Lordstown's IPO.[24] Burns admitted that a consulting firm generated nonbinding orders but said it was market research. "We are not stating these are orders and have never stated that," Burns said.[25]

That seemed to contradict statements he made during a CNBC interview four months earlier, when Burns said of Lordstown's preorders: "Most of them are signed by the CEOs of these large firms," and continued, "They're very serious orders."[26] Investors decided for themselves. The stock tanked 16.5 percent that day and never recovered. Lordstown filed for bankruptcy in June 2023.

. . .

Both GM and Hindenburg also were central players in perhaps the most spectacular EV flameout.

In June 2020, an electric-truck startup named Nikola (the last name of the Serbian American engineer and inventor, Nikola Tesla, was already taken) raised about $700 million in a SPAC deal. Nikola promised to unlock the potential of hydrogen fuel cells by becoming a one-stop shop for trucking companies to get hydrogen-based trucks on the road.[27] Hydrogen-fuel-cell technology has been in development for decades, with a smattering of commercial offerings by Toyota, Honda, and others, but hadn't caught on for a number of practical reasons, including the high cost and scarcity of refueling stations.

Nikola's idea was to sell not only hydrogen-based semi-trucks to commercial customers like beer maker Anheuser-Busch InBev, but also install and operate the charging stations, provide the fuel and related infrastructure, and so on.

There was reason to think that Nikola could be a serious contender. Compared to many EV startups, there were more grownups in the room. The executive behind the blank-check company that merged with Nikola was Steve Girsky, a former GM vice chairman and Morgan Stanley automotive analyst, well known in automotive circles. The company had attracted Germany's Bosch, one of the world's largest automotive suppliers, to invest in Nikola and provide its fuel cells. IVECO, a heavy-duty European truck brand, would build Nikola's trucks.

The counterpoint to this corporate roster was the brash young CEO of Nikola, Trevor Milton. In his late thirties when his startup burst onto the scene, Milton was an entrepreneur from Utah who had started a few companies, including one that made a fuel-saving system for trucks. Milton loved the spotlight and for a few years had been known for making outlandish claims about his various startups. In 2012, he said in a letter to investors that an e-commerce company he started "will become the largest competitor to Ebay in the next 1-2 years!" (It didn't).[28]

Nikola shares began trading on June 3, 2020, a Wednesday, at a stock market valuation of about $3.3 billion.[29] That weekend, Milton tweeted that Nikola would begin taking reservations for a *battery-powered* pickup truck, called the Badger. It was an unexpected disclosure. The company had its hands full developing the hydrogen-fuel-cell business. An EV pickup truck for retail buyers hadn't been seen as a priority.

Investors didn't seem to care. When trading resumed Monday, Nikola shares skyrocketed to $73, from $42. That vaulted its market valuation to $30 billion, higher than Ford at the time. Two days later, Milton doubled down: "My goal is to take the throne from the Ford F-150," he declared.[30] In other words, this startup that had yet to sell a single vehicle, and that few investors had ever heard of until a week earlier, was aiming to outgun the top-selling US model for forty straight years, despite the fact it went public on an entirely different premise of creating a hydrogen-fuel-cell trucking infrastructure.

Milton's provocations—along with a major deal with the nation's biggest automaker—grabbed more investors' attention. In September 2020, GM said it would help Nikola develop and build the Badger in exchange for an 11 percent ownership stake. GM shares popped. Nikola's popped even more. Steve Girsky, the former analyst and Queens native who had helped take Nikola public in the SPAC deal—and Mary Barra's former colleague—had connected the two executives.[31] Barra and Milton appeared together in a split-screen interview on CNBC, where Barra explained that GM would essentially be doing everything—providing the engineering, the batteries, the fuel cells, making the trucks. Milton struggled to articulate exactly what Nikola was bringing to the partnership. But he didn't lack superlatives. "The Badger is going to truly be one of the most amazing vehicles ever built," he said.[32]

Milton's hyperbole by then had landed on short-seller Hindenburg's radar. Started by Nate Anderson, a little-known investor with an investigative bent, Hindenburg was a specialist in forensic financial research. Two days after the GM-Nikola deal was announced, Hindenburg issued a sixty-seven-page report that was not coy in its conclusions: "Nikola is a massive fraud constructed on dozens of lies."

Among the most damning allegations: Nikola faked a promotional video showing one of its trucks cruising down a highway by actually rolling the inert rig down a long hill; Nikola—despite Milton having claimed that the company has produced hydrogen at a cost far below other suppliers—had in fact never produced hydrogen at all; and Nikola's backlog of fourteen thousand truck orders—or about $10 billion worth—was largely fluff, including one prospective customer that Nikola said was signed up for about $3.5 billion worth of trucks, but only had $1.3 million in cash to spend.[33]

Milton denied most of Hindenburg's claims, but the combative CEO and his startup would unravel quickly. Federal investigations began. Two weeks after Hindenburg's upbraiding of Nikola, the *Wall Street Journal* reported that BP and other possible partners to build Nikola's hydrogen-refueling stations had walked away from potential deals, sending the startup's shares down 25 percent.[34] Before the month was out, Milton stepped down. A federal jury eventually would convict Milton of securities fraud for lying about Nikola's technology.

GM of course pulled out of the major aspects of its Nikola deal. The lingering questions after the whole embarrassing episode, though, were "Why would a *Fortune* 20 company rush into the arms of an unproven startup and its sketchy founder? How did GM's army of lawyers and all their due diligence not uncover what a small group of some former journalists at Hindenburg had uncovered in a matter of weeks?"

That Hindenburg report—which largely stuck to the concrete details of its thorough investigation—did offer up some editorializing on this point. GM, it concluded, was looking for some Elon-esque buzz. "The real 'value' for GM seems to be branding," the Hindenburg report said. "We believe the legacy automaker simply seeks to latch General Motors' storied name onto Nikola's charismatic Founder."[35]

. . .

Barra and her team had had the Rivian opportunity snatched away by Ford two years earlier and now had suffered the high-profile Nikola blowup. There were mounting signs that even a company like GM—celebrated for decades for innovations like airbags, cherished models like the Corvette, and a breakthrough like the Chevy Volt—was struggling to manage the transition to EVs.

GM had promised investors it would launch its Hummer EV by the end of 2020. It kept that promise—sort of. The company ended up delivering one Hummer. Production was supposed to gradually rise in 2021. But GM finished that year having sold a paltry 854 of the hulking trucks. In June of 2022, GM was making just a dozen Hummers a day on average at its Factory Zero in Detroit. The Drive enthusiast website did the math: at that production rate, and 77,000-person deep waitlist, it would take seventeen years to fill GM's Hummer backlog.[36]

The next spring, media reports indicated that things hadn't progressed at Factory Zero from nine months earlier: still only around a dozen trucks a day on average were trickling out of the plant.[37]

GM was crawling into its EV transition. But why? Hadn't it made tens of millions of cars over a century? Yes, it had the car-making part down. Its workers knew how to bolt seats into the floor. But there were major differences in building EVs, especially in the supply chain. If an automaker doesn't have a steady supply of battery cells, it can't ramp up its EV production. In 2022, the year GM had been promising investors an onslaught of EVs, the company had anything but a steady supply of batteries.

The people inside GM's vast and powerful purchasing division weren't accustomed to being told no. For most of GM's history, it was so powerful that it could strong-arm just about any supplier into submission. But it and other traditional automakers held no special sway with battery suppliers, and there were only a few of them with serious scale on the planet.

GM was trying to play catch-up, but it couldn't. For all its early talk about going big on EVs, it went too small. The production volumes it planned back in 2018–2019 didn't foresee the surge in consumer interest that began after the pandemic. Once demand took off, Barra and her team tried to course-correct by increasing production plans for the Cadillac Lyriq, the Chevy Equinox, and other new EVs. But there weren't enough batteries to go around.[38] Companies like Panasonic, LG, and Samsung were tapped out through the first half of the decade. And GM's own battery factories were still under construction. Once GM finally did have enough battery cells, an automated factory system tasked with bundling them together struggled for months to work properly, further bogging down EV launches.[39]

It was becoming clear to investors that the quest for the next Tesla might be a futile one. By late 2022, Tesla was a juggernaut, occupying a space between traditional autos and all the EV wannabes. Seemingly overnight, Tesla had become more profitable than those legacy carmakers that for so long had dismissed it as a money-losing fantasy that could never master the art of car making. And yet Tesla was still acting like a fast-growing startup; sales surged 40 percent that year.

But even for Elon Musk's company, that journey from startup to large-scale manufacturing was harrowing. During Tesla's infamous period of "production hell," from 2017 into 2019, as it made the leap to producing hundreds of thousands of EVs annually, the company at one point was about a month away from running out of cash, Musk has said.[40] That was more than fifteen years after Tesla's founding.

"Prototypes are easy. Production is hard," Musk would say in the fall of 2023, when Wall Street analysts were pressing him for details on when his much-hyped and long-delayed Cybertruck—the stainless steel pickup that would look comfortable on the set of a *Mad Max* movie—would actually start rolling out in real volumes. His answer probably didn't thrill investors. It's "like 10,000% harder to get to volume production than to make a prototype," he said. "And then it is even harder than that to reach positive cash flow. That is why there have not been new car startups that have been successful for 100 years, apart from Tesla."[41]

That's an ominous take for the EV newcomers like Scaringe's Rivian, still struggling to scale up. But here's the rub: EV upstarts in the early 2020s would have it even harder than Tesla, because they faced some additional huge hurdles that Elon never had to worry about.

One is competition. Outside of China, Tesla essentially had a blue ocean, a market to itself for a decade. Rivian, Lucid, and others face an influx of new models from well-capitalized traditional automakers that finally bought in on EVs, even if their early efforts stumbled. Another difference: by the time these latest EV startups were in launch mode, the era of cheap capital was over, with governments around the world raising interest rates to combat post-pandemic inflation.

And these companies were trying to launch cars in a most dysfunctional supply-chain environment. From computer chips to battery cells to even mundane items like seat foam, parts shortages were disrupting automotive factories worldwide. Despite the onslaught of investor giddiness, the timing of firing up an EV factory in the early 2020s, as Scaringe would say later, "was almost perfectly bad."[42]

· · ·

Back in Normal, Illinois, the factory frustrations were there from the start. It turned out that much of the equipment Rivian's team hoped to salvage from the Mitsubishi operation wouldn't work. The giant stamping presses were salvaged, but not much else. Dozens of semi-trucks packed with factory equipment began arriving early in 2020, but by mid-March, the country went into pandemic lockdown. Rivian initially didn't have enough workers to empty the trailers, let alone to install the huge steel machinery and other gear on the factory floor. Pandemic restrictions prevented Rivian engineers from flying to Normal from California to help with factory prep.

At this point, early prototypes of the pickup trucks were being hand-built off-site and had to endure rigorous testing. In March 2020, Scaringe and his team anticipated the buildings that housed the testing computers were going to be shut down, so they spent some forty straight hours wheeling equipment out of the building so they could get it hauled to engineers' homes and set up in their garages. Eventually, the company came up with a color-coded system for the factory, where engineers and contractors were split into four groups to prevent comingling in case of a Covid-19 outbreak. By that summer, some 1,800 workers were swarming the Normal plant daily, fitted with hard hats, safety glasses, and masks.

Scaringe had been living in California, but that wasn't going to work. He moved along with his wife, Meagan, and three young boys near Normal.[43] The CEO's impulse to be on the ground laid bare his hands-on management style.

"He's brilliant. He gets stuff done. He can solve any problem," said one executive who worked closely with Scaringe. "But also, he's a perfectionist. He's pretty obsessive."[44]

In 2020, prior to Scaringe's move to Illinois, Rivian's head of software strategy, an Apple recruit, left the company. Scaringe, who had limited hands-on experience in software development, relocated his desk to the software division at one of Rivian's buildings in California and spent months delving deep into strategy and product development.

In Normal, Scaringe arrived at the factory one day and decided he wasn't happy with the shade of white paint that contractors used on the factory's beams and pillars. He ordered them repainted, a $1 million job, before his team talked him out of it. Another time, the CEO didn't like the shade of paint used on some of the factory's robots. This time, they couldn't talk him out of his decision to repaint them in a slightly different hue.

During the production launch, bottlenecks in the body shop section of the assembly area slowed progress. So, one weekend, Scaringe moved his desk into the body shop, just a few feet from bustling factory workers. On Monday, some on his executive team tried to gently tell Scaringe that having the CEO plopped in the middle of the body shop wasn't the best way to troubleshoot.

"I'm a good problem-solver," Scaringe responded. "I'm down here solving our biggest problem." Scaringe would pull the desk move several times during the early years at Normal, sometimes stationed for weeks at a time amid the clattering work of the line.[45]

For all Rivian's troubles, getting parts was the most vexing. Every morning, Scaringe would receive an email detailing the most problematic suppliers. It came in the form of an "escalation list"—a series of steps the team would take to pressure the supplier to deliver the parts on time. The CEO would start his day by scanning for the suppliers most risk of shutting things down, and then phone that company's CEO.

"This was just like, a bare-knuckled fight to get parts," Scaringe would say later.[46]

All the other automakers were waging the same fight. Computer chips were the biggest bottleneck, but other random shortages would crop up almost every day. A startup like Rivian had very little leverage, but very much to lose. Rivian might only need 50,000 units of the same part that GM and Ford need 8 million of. The supplier prioritizes volume. But for Scaringe, not getting that one little widget could shut down his only factory. And it often did.

"You'd walk through the plant, and it was like, 'Oh my God, nothing is happening,'" Scaringe would recall later.

This scenario was playing out in triplicate, across all three Rivian vehicles. Scaringe later would acknowledge that trying to do both Rivian vehicles plus the Amazon van was overly ambitious, but said he couldn't have predicted the chaos of the pandemic-related aftermath. "We made our bed, and then we had to lay in it," he said.[47]

By the spring of 2022, the factory floor at Rivian was a frenetic mess. Boxes and bins of parts would routinely stack up throughout the factory. Despite Scaringe's haranguing of supplier CEOs, parts wouldn't arrive on time and began to back up. Lots of parts. The assembly line often stopped, sometimes for hours. Such interruptions are extremely rare at a typical car plant—millions of dollars in revenue can be erased for every hour of downtime. The parts shortages weren't just killing production, but causing a bigger mess inside the plant, as other components kept streaming in, causing a colossal backup.

Finding places to stash all those bolts, steering columns, bits of interior trim, and thousands of other parts sucked up gobs of time and resources throughout 2022. Rivian grabbed space in an old furniture warehouse across the street. Eventually, it would have leases on eight parts warehouses spanning from south suburban Chicago to Missouri.

The parts backup also scrambled the intricate staging process needed to ensure components get to where they're needed on the assembly line at the precise time. For a green pickup truck, the staging area alongside the line needs to have bins with a green front fascia, a green door handle, a green mirror cap, and so on. That process, called parts sequencing, is complex, especially at a new factory making first-generation models. Scatter those parts across eight warehouses through a several-hundred-square-mile area, and you've got a nightmare puzzle to solve.

Problems were evident outside the factory, too. Even when parts shortages stopped the line, multiple delivery trucks kept arriving every day with parts from the various warehouses. But without parts going out to the assembly line and into cars, they would run out of places to put the stuff coming off the trucks.

At one point, more than seventy semi-trucks were staged in a snaking line around Rivian's campus, waiting hours or even days to offload. "All the truck drivers were pissed off," Scaringe recalled later. To relieve tensions, managers doubled the normal daily cafeteria order to offer the truck drivers sweet-potato hash and other artisan Rivian meals.[48]

As these disruptions played out in Normal, there was building consumer frenzy for electric vehicles throughout the year. For an all-EV company, that's a good thing. But there were complicating side effects. The laws of supply and demand sent prices for battery materials like lithium and nickel soaring. Those extra costs would make the already substantial losses on every Rivian much bigger. In the first part of 2022, carmakers began raising prices on EVs to offset the surge in materials costs.

Rivian announced in March that it was raising prices on its trucks by as much as $15,000. But not to worry, the company said: any customer who wasn't willing to eat that increase could always wait two more years and get a cheaper, less-powerful truck with less range. Rivian cited "inflationary pressure on the cost of supplier components and raw materials across the world."[49]

On paper, the move might have been justified. Lithium prices alone had more than doubled in a matter of months. Prices for sheet metal, semiconductors, and other components had shot up too. And the cost of inflation spanned beyond EVs. The post-pandemic vehicle shortage had created the mother of all sellers' markets. Dealers were jacking up prices well above sticker, seeing their new-vehicle profits grow threefold. Companies like GM, VW, and Toyota were profiting handsomely too. If those mature rivals were enjoying this pricing tailwind, why shouldn't Rivian?

There was, of course, a big difference. Rivian wasn't just raising prices on anonymous future car buyers who might wander into a showroom. The tens of thousands of reservation holders who had plunked down $1,000 reservations years earlier were the company's de facto brand ambassadors. They were the people out there showing screenshots of their future R1T to neighbors, spreading the gospel of this outdoorsy, American EV-truck maker with the cool factor of Tesla, but with a softer, Patagonia vibe.

The backlash was instant and harsh.

"I don't know if I've ever gone from feeling so good about a company to so bad," posted one Reddit user, whose R1S went from $76,000 to above $90,000.

At Rivian's headquarters in Irvine, executives were scrambling within an hour of pushing out the news. "It had a very West Wing feel to it. A lot of walk-and-talk discussions and phone calls and people calling huddle meetings," said one manager who was involved. "It wasn't whether we would walk it back, but how."[50]

Two days later, Scaringe released an apology letter and said Rivian would stand by existing preorders at their original prices. "It was wrong, and we broke your trust in Rivian," he said.

Those price hikes would stick for future customers, though. Rivian would need every bit of that as it sought to erase huge, persistent losses. But what if, by the time Rivian worked through that backlog of preorders, the frothy pricing on EVs disappeared? What if all these EV makers finally got over the hump and started cranking out more EVs, just in time for their heady pricing power to evaporate? Elon Musk, yet again, was about to scramble the industry's best-laid plans.

15

An EV for Everyone?

n the summer of 2023, Chris Lemley's car dealership near Boston got a call from Ford corporate. Great news, the field rep said: after a year of constraints on the F-150 Lightning pickup, the factory was now crank-ing. Ford could ship more than twenty of the electric trucks to Lemley's store, just up the Mystic River from Boston Harbor.

Six months earlier, Lemley's sales staff would have been elated. Like many Ford dealers across the country, his dealership had kept a long wait list of Lightning buyers who seemed unfazed by the $80,000 to $90,000 price; they just wanted to be the first in their neighborhood to have the truck.

By that summer, though, there had been a noticeable shift in the type of customers who came in to kick the tires on a Lightning. They weren't any longer the early adopters willing to pay whatever it took. There were more questions about the price and the monthly payment—questions for which Lemley's staff had no good answers. The average lease payment was around $1,700 a month, he said, and some were over two grand. "A lot of my salespeople were afraid to quote the lease payment," Lemley said. "They didn't want to look like crooks."[1]

By this point, many car dealers had lost their enthusiasm for electric cars. EVs already had been a high-maintenance sale. Staffers had to spend extra time with potential buyers compared to gas-powered cars, explaining nuances like charger installation and battery-health reports. And all that effort was going to sell a vehicle that wouldn't bring much money back to the service departments either.

During the years of the EV euphoria—roughly from 2020 through early 2023—many dealers were willing to overlook these costs and hassles. They had long reservation lists and were reaping extra profit from frenzied buyers paying above sticker price. Now, many had unsold EVs stacking up on their lots. Electrics were taking an average of 103 days to sell, more than double that of internal-combustion vehicles—a full reversal from a year earlier when EVs were flying off dealer lots faster than they could get them.[2]

For dealers, EVs had whipsawed from a buzzy source of sales and profit to a capital P Problem. Many dealers were straight up opposed to President Biden's EV push on ideological or political grounds as well. A mini-revolt was fomenting.

Lemley wasn't your typical car dealer. He had graduated with a degree in American Civilization from Brown and got an MBA from Harvard.[3] His store is located in a liberal hotbed, a stone's throw from Cambridge and Boston. Lemley was far from a knee-jerk-reaction, anti-EV guy. But when he heard that Ford was looking to drop off a few dozen Lightnings at his dealership that summer, he politely declined. "I think this was a conversation they were having with every dealer in the country," Lemley recalled later. "And just about every dealer in the country choked on the number."[4]

It's not that Ford was oblivious to the EV market's shifting dynamics. Ford executives were at this time among the first to signal to investors that their electric plans weren't playing out as expected. Buyers weren't willing to spend the same hefty sums that people had been paying even a few months earlier. This was going to take longer than the car companies had thought.

Later that summer, Ford CEO Jim Farley was on his photo-op-friendly road trip through California. In between showing off EVs to Dwayne Johnson, blowing up bouncy houses with his Lightning, and doing doughnuts in the Walmart parking lot, he was on a reconnaissance mission of sorts to take the pulse in the nation's biggest EV market. One Ford dealership's billboard along Interstate 405 near Los Angeles advertising the Mach-E seemed to practically taunt Farley: "100+ in stock & available now! We're making deals!"

This flip happened shockingly fast. "My number one question I've been asking every dealer on this trip," Farley said, "is 'What the hell happened in the last ninety days?'"

He was asked by a reporter whether, after all the plans and press releases and money being steered into the EV transition, what if it just peters out? What if it doesn't take off?

"I ask that question every day myself," he said. "I'm putting $50 billion into this."[5]

This sudden cooling in consumer demand was scrambling the narrative that many car executives and policy makers had spent years constructing: that the budding market for electric cars was going to surge through the 2020s, and that EVs could account for as much as 50 percent of vehicle sales by the end of the decade.

That scenario seemed possible during the wave of enthusiasm in the wake of the pandemic. Within the span of a few years, hundreds of billions of dollars in capital were earmarked to EV development. Designers and engineers went into overdrive to get EV models out sooner than planned. Excavators began clawing into the earth on dozens of massive construction projects, like Ford's battery factories in Kentucky and Tennessee.

The conventional wisdom was that EV sales had hit the hockey-stick growth curve—after a slow, steady climb, they would rocket higher for years. There were expected to be more than a hundred EVs in US showrooms by 2025.[6] Industry executives talked themselves into believing that the expense of producing an electric car would roughly match that of internal-combustion vehicles by mid-decade as battery costs fell. Federal subsidies would continue to be a carrot for consumers to try electric. And the billions in federal money pouring into new charging stations would begin to quell worries about finding enough places to power up.

Just months before Farley's California road trip, Ford was tearing up fresh concrete to expand its Lightning plant in Detroit for a second time in six months.[7] GM was still straining to get its new EVs out, with CEO Barra promising a "breakthrough" year for its electric sales in 2023.[8] Volkswagen was touting an order backlog of 350,000 EVs in Europe alone.[9]

"With EVs, right now it's like, 'You build it, and they come,'" Steven Center, a US Kia executive, said that year. At the time there was a wait of up to six months for the Korean brand's sporty EV6.[10]

But even then, in the hysteria, tiny red flags were waving. And Elon Musk, ever on brand, was the one waving them.

. . .

One rainy Saturday afternoon in Los Angeles, Stanly Tran was online browsing car-shopping sites and social-media posts when he should have been working on his dissertation for his doctorate in clinical psychology. In his early thirties, Tran, a psychotherapist, had spent several months researching EVs. It was January 2023, and he wanted to upgrade from his small Chevy Bolt EV before he and his wife had their first child that spring. There were several new electric models on Tran's radar: the Ford Mustang Mach-E, the Kia EV6, Hyundai Ioniq 5, and Tesla's Model Y.

That day, Tran came across some chatter on Reddit about a massive price drop on the Model Y. Tran's jaw dropped: The sticker had gone down by more than $10,000. He hopped over to Tesla's site. Sure enough, the Model Y that he had been eyeing—a long-range version with all-wheel drive—had dropped by around $14,000 from the last time he had checked.

"This is a no-brainer," Tran thought. He quickly called Tesla and got a customer-service rep to assure him that the price was correct. Tran plunked down a deposit. He told his wife as soon as she returned home. "What about the Mach-E?" she asked. Tran explained that the value on the Model Y was too good to pass up.[11]

Tran's screamin' deal was among the first of what would be a series of price cuts from Tesla in early 2023. Given Tesla accounted for more than 60 percent of US EV sales, its moves triggered a chain reaction. People in search of an electric, like Tran, were pressed into quick decisions. Used-car dealers who had Model 3s and Model Ys in stock saw their valuations plummet by thousands of dollars overnight. Customers who had bought a Tesla in the weeks and months before the cut were furious. "Hey Elon Musk and Tesla, not all of us are billionaires," one owner posted on social media, complaining that the same Model Y she bought for $72,440 a few months earlier was going for $59,630.[12]

Tesla investors were a bit agitated, too, wondering what these deep cuts meant for Tesla's vaunted profit margin. And more importantly, what did this say about demand for Tesla's cars? Since the company's "production hell" period five years before, the underlying question had been about supply: could Tesla crank out enough cars to meet the accelerating demand for EVs in the United States, Europe, and China? But now these deep rollbacks were raising questions about whether the market for Teslas—and, by extension, for EVs broadly—had plateaued. Was Tesla's long-term goal of logging 50 percent sales increases for years to come still a viable one? Was the EV thing just a fad?

Days after Tran clicked the "purchase" button on his Model Y, Elon Musk dialed into a highly anticipated conference call to discuss Tesla's earnings. The dozens of analysts at Wall Street's biggest banks who cover every twist and turn of the Tesla story wanted to know, above all else, what the deal was with these price cuts.

Musk began the call in a chipper mood, noting that 2022 was a record year for Tesla in almost every aspect: vehicle sales of 1.3 million; a 17 percent operating margin that blew away other automakers; and net income of $12.5 billion.

But these analysts already knew Tesla had a good year. They were more worried about the coming years. Was Tesla slashing prices because not enough people wanted to buy its cars?

"I want to put that concern to rest," Musk told them. He said the company was logging vehicle orders at almost twice the rate it could produce them. The analysts gently pushed back, trying not to trigger Musk's notorious temper: of course demand is up, because you just cut prices 20 percent. But won't these fire-sale prices pummel your profit margins?

Musk pointed out that costs had fallen, as prices for lithium and other battery raw materials fell. Then he got a bit philosophical. For two decades, he had talked about democratizing EVs. He wanted to spread the technology to as many people as possible to help the planet. These lower prices, he explained, didn't mean Tesla was no longer a hot brand. He just wanted to broaden the fan base.

"I think there's just a vast number of people that want to buy a Tesla car but can't afford it. And so these price changes really make a difference for the average consumer," Musk said on the call. The average Tesla customer—still overwhelmingly affluent men—may not think about affordability, he explained. But "it's always been our goal at Tesla to make cars that are affordable to as many people as possible."

Later in the call, Musk suggested another reason that investors should look past the lower prices: Teslas, he explained, come with embedded hardware that allow the car to take over some aspects of driving, such as automatic steering or changing lanes. Eventually, he said, Tesla would be able to offer a simple software update that would make millions of Tesla cars fully self-driving. "That might be the biggest asset value increase of anything in history." Translation: Musk was relatively unconcerned about the selling price because he believed nearly every Tesla on the road could produce vastly more revenue in the future—a profit bonanza with the flip of a switch.[13]

Keep in mind that at this point, few if any experts believed that Tesla was anywhere near fully autonomous capability. But the upshot for car executives in Detroit, Tokyo, or Germany was simple: it didn't really matter whether Elon Musk was right or not. Regardless of his motivation, his moves amounted to a giant blaring alarm. The biggest player in the EV game had just started a price war at the worst possible time, just as these legacy companies were straining to squeeze out every gram of cost in a bid to claw their way to meager profit margins. Musk could have said he was doing this because he thought unicorns would shoot out of the sunroof—it didn't matter. His behavior was creating a new, uncomfortable reality for some of the world's biggest companies.

"It's safe to say this was an oh-shit moment for all of us," a GM executive said later.[14]

Everyone knew that the traditional car companies were still losing money on EVs. Despite all the efforts being put toward driving down costs—bringing batteries in-house, introducing flexible platforms that could share mechanical components, tweaks in battery chemistries—the path to profits for VW, GM, and Ford was still riddled with potholes. It had been such a strong seller's market since 2020 that pricing pressure had been the furthest thing from these executives' minds. Many customers had paid deposits for multiple EVs—a Rivian, a Ford Lightning, and Chevy Silverado EV—and were determined to buy whichever they could land first.

But those buyers' patience was wearing thin just as Musk began slashing prices. Daniel Heinzen was one of them. Heinzen, a Chicago construction manager in his mid-thirties, owned a GM-built Cadillac Escalade. The massive SUV is a staple choice among professional athletes and celebrities, but is among the most fuel-thirsty vehicles available anywhere. A marketing email from Cadillac in 2022 alerted Heinzen to the pending arrival of the 120-year-old brand's first fully electric model: the Cadillac Lyriq, a sculpted midsize SUV. Heinzen began thinking the Lyriq could be a good option for his wife. The couple had a baby, and her daily routine generally was confined to trips around the city—just the type of driving that drained the most gas. He liked the idea of her bypassing stops at the gas station. He knew a few people who drove Teslas and liked the design and the tech features. But he was turned off by the interior, which he thought was plain. He had seen reviews online praising the Lyriq's high-end interior. "I don't really view Tesla as luxury," Heinzen said. "When I think of American luxury, I think of Cadillac."

He put down a few hundred bucks for a reservation to buy a $63,000 slate-gray Lyriq. Heinzen figured it would take longer than a year, but he was willing to wait.

That began a more than two-year headache. Heinzen would go months without updates from his dealer. Nearly a year after he put down his deposit, GM emailed him: the 2023 model year had run out. If he still wanted a Lyriq he would need to switch his order to a 2024 model—and absorb a price increase of around $13,000, pushing his tab to about $76,000. Heinzen, who served as a soldier in Iraq during the war in 2005, was bummed about the price increase, but he still wanted one and he made the update to his order.

Heinzen eventually got tired of waiting, so he bought a used 2018 Tesla Model 3 for about $27,000 as a kind of dry run to see if they could live with an EV. The couple loved it, and with a second baby on the way, they still wanted to swap it out for the Lyriq. Months dragged on, and still nothing from his dealer or from Cadillac. So instead Heinzen bought a lightly used 2022 Tesla Model Y, for $39,000.

"Teslas are everywhere in Chicago," he said. "I eventually just decided that they can't be all bad."[15]

For the legacy car companies, this state of play was scary. Some, like GM, were still struggling to ramp up production of their new EVs, and they needed more scale to make them profitable. But now the market was sending mixed signals on consumers' enthusiasm. Meanwhile, Tesla was still making plenty of money, even at discounted prices. It had a proven brand and a devout following. It had established itself as a major player in China and Europe.

GM, VW, and Ford had none of those. Ford—the only company at this point to disclose the financial performance for its EV business—lost about $4.5 billion on them in 2023. Tesla had pretty much soaked up the early adopters—the tech bros, engineers, and environmentalists who were super-motivated to drive electric.

The market for EVs still was growing at a faster pace than the rest of the market. EV sales in 2023 rose about 47 percent from a year earlier, compared to 12 percent for the overall market. But that EV sales pace was down from nearly 70 percent the year before. This is in some ways statistically inevitable. It's harder to double a million sales than it is to double a hundred thousand. But the pace was slower than most auto executives had counted on when they set all those bullish EV sales forecasts, and they were rattled.

EV sales had shot up from almost nothing to about 8 percent of the US market over five years. Many executives and experts had expected a jump to 15 percent or 20 percent would happen just as rapidly, as costs edged down, driving ranges got longer, and more chargers were installed, soothing buyers' concerns about charging.

There was also a slew of new models coming to give buyers more choice. Several fully electric three-row SUVs were headed to US showrooms—

including a battery-powered version of that brutish, gas-guzzling Escalade—finally putting EVs in one of the fastest-growing and most profitable markets.

There were a lot of reasons why EV sales should have been poised to take off. So how did a market that had been so red-hot cool off so quickly? Why were auto executives suddenly worried?

It wasn't necessarily customers' lack of interest. Plenty of people were intrigued by the idea of ditching internal-combustion cars and avoiding gas stations. In a spring 2023 survey by research firm Cox Automotive, more than half of respondents said they would consider an EV for their next purchase, up from 38 percent a year earlier.[16] This prospective buyer base, though, was more discerning than the true believers who had helped drive Tesla's growth for a decade. Those fervent customers were so committed to the notion of driving electric—and the Tesla brand—that they tended to overlook the inconveniences of EV ownership. For most potential customers, switching to electric is a significant lifestyle change. The idea of plugging your car into the wall of your garage every night takes a mindset shift (and typically an electric-panel upgrade). So does having to map out your next road trip and set aside a few more hours to figure out places to charge.

Just about any survey from this period found that unease and uncertainty around range anxiety and charger availability were at or near the top of the reasons people refrained from going electric. Most of the new EVs hitting the market were coming from companies that, unlike Tesla, didn't have their own charging networks. Many of those car brands were scrambling to offer their customers some access to Tesla's walled-garden Supercharger stations, but that was still at least a year off. So they were left trying to convince consumers that their new electric models were worth taking a chance on.

Another impediment to adoption cropped up in 2023: the EV movement became a pawn in the culture wars. Electric cars had been an easy target for conservatives for years. But two factors emerged in the early 2020s that drew a political and ideological backlash: the Biden administration's spending on climate initiatives; and the United States' increasingly hawkish view of China, the world's EV hub.

The Biden-backed Inflation Reduction Act—the massive climate-spending package that became law in late 2022—passed with 270 senators and House members voting in favor. Not a single yea was cast by a Republican.[17] Conservatives saw the wave of federal spending on climate initiatives—a half-trillion dollars by some measures—as market manipulation at best and, at worst, an un-American fleecing of civil liberties.

"They are still trying to shove this down people's throats," said US Representative Sam Graves, a Missouri Republican and chairman of the House Transportation Committee.[18]

A survey in July 2023 by polling firm Morning Consult found that nearly one in five respondents viewed EVs unfavorably for political reasons. And a whopping 60 percent cited the China connection as a reason they were down on EVs.[19] (Never mind that many parts in those respondents' gas-engine parts have Chinese origins as well.)

Automakers were caught in the middle. The political backlash became yet one more thing standing in the way of EV adoption. There were millions of Americans who wouldn't drive an EV, even if they had an ample budget to buy one and a massive garage in which to charge it.

"I never thought I would see the day when our products were so heavily politicized, but they are," Bill Ford, told the *New York Times* in October 2023.[20]

Stepping into this heated debate were thousands of US car retailers. It's hardly a secret that dealers tilt conservative. They are, after all, entrepreneurs, leery of the higher taxes and red tape that can complicate their businesses. During the week after Thanksgiving in 2023, nearly four thousand dealers signed an open letter to President Biden, warning him that his EV agenda was out of step with consumer sentiment.

"Last year, there was a lot of hope and hype about EV. But that enthusiasm has stalled," the dealers wrote. EVs, the letter said, "are stacking up on our lots."

The dealers' specific target was the president's proposed fuel-economy regulations, which were expected to be finalized the following spring. The rules floated by the EPA didn't mandate that Americans buy EVs, but they might as well have, the dealers argued. The regs would require such a big reduction in tailpipe emissions that, effectively, more than half of vehicle sales by 2032 would need to be all-electric for carmakers to comply. Even the most bullish pundits weren't projecting EV sales to hit that level by then.

Most consumers, the dealers told the president, were simply not ready to make the switch. They worried about charger availability, sure. But dealers were on the front lines to hear about the other specific hang-ups consumers had with the idea of going electric. They worried about the very real problem of a loss of driving range in hot or cold weather. Truck buyers didn't think they were any good for towing a boat or horse trailer, the dealers said.

"Mr. President, it is time to tap the brakes on the unrealistic government electric vehicle mandate. Allow time for the battery technology to advance. Allow time to make [battery-electric vehicles] more affordable. Allow time to develop domestic sources for the minerals to make batteries. Allow time for the charging infrastructure to be built and prove reliable. And most of all, allow time for the American consumer to get comfortable with the technology," the dealer letter said.[21]

. . .

Around this same time, auto executives used their quarterly conference calls to update investors on the state of play on their EV plans. It was a jarring reversal from the message of previous quarters, exposing deep angst around their expectations for how fast consumers were willing to switch to EVs.

GM's Mary Barra greeted investors by walking back a key EV goal. GM had been aiming to produce 400,000 EVs in North America over a two-year period, ending in mid-2024. It was an arbitrary figure. But it was significant because Barra had set that bogey to stretch her team and prove that GM was going to be a force in the early EV race. And Barra had become known among investors for delivering on her promises. Over the previous nine years, she had beaten Wall Street's quarterly profit estimates in thirty-four of thirty-five quarters, a remarkably consistent track record.[22]

Some of these EV projects that GM had been hurrying to pull ahead would now be pushed back. A $4 billion gut overhaul of a suburban Detroit factory to make EV pickup trucks would be delayed a year, to 2025. GM also would push off the introduction of a few new electric models.

"As we get further into the transformation to EV, it's a bit bumpy, which is not unexpected," Barra told analysts.[23]

Two days later and ten miles away in Dearborn, Ford CEO Jim Farley and his team were even more candid about the need to pump the brakes on the race to EVs. The company said it would delay about $12 billion in planned spending on EVs.

"Things are changing. EVs are still in high demand," but, he said, "the pricing is much lower. And there's a lot of overcapacity." A year earlier, the companies had been trying to convince investors that they could make the switch to EVs the fastest. Now, it was almost as if they were making the case that they were best positioned to slow-walk the transition.

"Ford is able to balance production of gas, hybrid and electric vehicles to match the speed of EV adoption in a way that others can't," Ford CFO John Lawler assured investors.[24]

The slowdown in EV sales began cascading through the supply base, too. By late 2023, the parking lot at Our Next Energy's headquarters outside Detroit was no longer overflowing onto the lawn. The battery startup on the Monday after Thanksgiving weekend laid off a quarter of its staff, or about 130 people, citing "market conditions."[25] Founder Mujeeb Ijaz was demoted, from CEO to chief technology officer, although his dream of making batteries in the United States remained alive—the Michigan factory project was still on.

Even Elon Musk sounded uncharacteristically downbeat and unsure of where things were headed. He refused to say whether a long-rumored Mexico factory was a go, because of concerns about how soon it would be needed. And he worried about how higher interest rates would prevent many would-be buyers from becoming Tesla owners. "If our car costs the same as a RAV4, nobody would buy a RAV4," he said.[26]

With that unprovoked shot at Toyota, Musk crystalized a point that all the auto executives seemed to agree on: EV prices were too high for mainstream buyers. Or at least too high when combined with those other reasons that made buyers hesitant, like range anxiety and charging accommodations. A large global survey from S&P Global Mobility in the fall of 2023 showed nearly half of those polled thought EVs were too expensive.[27] In North America, there were almost no EVs offered for less than $40,000.

The problem was, very few car executives seemed to be doing much outwardly to solve the affordability problem.

Barra earlier in 2023 killed GM's only affordable EV for sale in the United States, the Chevy Bolt. Later, GM said the customer backlash was so strong that the company decided to develop another one, but that would take a few years. GM was doing more than most rivals to move down the price ladder—a sharp-looking, battery-powered version of its Equinox compact SUV was to start at $35,000. But investors and car shoppers by this point had grown tired of GM's EV promises—they'd believe it when they saw these cheaper models in showrooms.

Farley often talked passionately about squeezing costs out of Ford's future EVs, but that didn't mean they'd be cheap. On the same call when he talked about weakening consumer demand, Farley gushed about a new electric pickup truck that his tech chief Doug Field was overseeing. He called the future pickup "one of the most thrilling vehicles I have ever seen in my career," with "a digital experience that totally is immersive and personalized." Sounds cool, but not a $30,000 EV that would bring electric driving to the masses.

By early 2024, Farley wanted to signal to investors that he had a plan for a lower-priced EV that could help Ford punch through to real sales numbers. The company had secretly assembled a skunk-works team of engineers to put together a small-EV platform that would serve as the foundation for many electric models. Farley wouldn't get into specifics but was candid about his motivations: "Our EV teams are ruthlessly focused on cost and efficiency," he told analysts, "because the ultimate competition is going to be the affordable Tesla and the Chinese."[28]

Musk still seemed more preoccupied with the pricey end of the market, rather than the $25,000 car he had been promising for years. On that October earnings call, Musk addressed both: yes, Tesla's EV for the everyman is still under development, he told investors. Tesla would deliver this via a breakthrough in manufacturing that would automate much of the process, leading to unprecedented cost reductions. But he declined to provide a timeline. And he warned people not to get too excited. "It's utilitarian. It's not meant to fill you with awe and magic. It can get you from A to B."[29]

A few weeks later, Musk appeared at Tesla's Austin, Texas, headquarters to meet some of the first Cybertruck customers and sign the hoods of their pickups. The space-age-looking truck with the bulletproof steel panels and lightning-quick acceleration was revolutionary, but hardly the thing that would solve the affordability problem. It would start around $61,000—about 50 percent higher than Musk had promised four years earlier—and top out around $100,000.

Even RJ Scaringe at upscale Rivian was feeling the need to appeal to mainstream buyers. For years, he had promised a smaller, more affordable truck. On a sunny Thursday in March 2024, inside a restored historic movie theater in Laguna Beach, the dapper CEO finally delivered on that promise by revealing a miniaturized version of Rivian's original R1S SUV. Cheers erupted when he announced the price of the EV, named R2: $45,000, to arrive in 2026. But Scaringe wasn't done. He surprised the crowd by beckoning onto the stage an even smaller, mini SUV, named R3, that promised to democratize the brand beyond affluent Patagonia wearers. Scaringe wouldn't peg a price though or say when it would go on sale.[30]

But there was one company that had already figured out how to bring compelling, low-priced EVs to the masses: BYD. Wang Chuanfu's company with the sappy "Build Your Dreams" slogan offered more than a dozen electric models, with fun names like Dolphin, Song, and Sea Lion, compared with Tesla's spartan lineup of four nameplates. One, the Seagull, started below $10,000. Another, the Seal, was in the high $50,000s and competed against with Tesla's Model Y. BYD was even eyeing the high end of the market too, with plans for a $233,000 supercar.[31]

By early 2024, BYD had put the car world on notice. The company Elon Musk had mocked on live television posted fourth-quarter sales figures that topped Tesla to earn the title of world's largest electric-car maker. For the traditional car companies, BYD was no longer a problem confined to China. The company was expanding into Europe.

On a Monday in February 2024, news crews gathered at Germany's port of Bremerhaven on the North Sea to watch as more than three thousand cars rolled off a massive blue-and-white cargo ship emblazoned with the BYD logo. It was the maiden voyage for the *BYD Explorer No. 1*, the Chinese automaker's first chartered vessel. Another seven car-carrying ships—with the capacity to hold up to seven thousand vehicles each—already had been ordered. BYD was now cranking out so many cars at its Chinese factories that it had begun ramping up exports to Australia, Brazil, Israel, and other markets. Europe was next, with plans to open a factory in Hungary by 2026, which would make BYD the first major Chinese automaker producing cars on European soil.[32]

Auto execs at the global carmakers feared North America would be next. In a rare video interview, Stella Li—BYD's number two executive, who two decades earlier had snuck battery-cell samples into the world headquarters of cell-phone companies—confirmed BYD's interest in building an assembly plant in Mexico. That would simply serve the Mexican market—BYD had no plans to enter the United States, Li explained. But the read-through for US car executives was painfully plain: a BYD plant in Mexico would give the company a beachhead for US expansion, because US trade laws could allow the cars to stream across the border without the hefty tariffs slapped on Chinese-built vehicles.[33]

If that thought wasn't chilling enough for competitors, the interviewer finished the conversation by asking Li what she thought of the recent moves by some global automakers to pull back on EV investments because of the murky demand picture.

"If you are not investing for electric car, you are out. You will die," she said. "You have no future."

Crosscurrents

f you're looking for a tidy, conclusive ending to this book, prepare to be disappointed. As we hit the mid-2020s, the electric-car story is as messy as ever. The popular industry prediction that EVs would hit a tipping point by now has not come to fruition. Except in the places that it has.

In China, the transition is cranking. EVs and plug-in hybrids run at around 40 percent of total new-vehicle sales. The United States remains a stubborn laggard at around 10 percent—except in California, the trendsetting, most populous state, where battery-powered cars account for more than a quarter of vehicle purchases. EV momentum in Europe seems to have stalled around 20 percent overall, but it's wildly variable.[1] In Norway, for example, the share of zero-emissions vehicles just keeps rising, zooming past 90 percent of the market in some months.[2]

Carmakers that had doubled down on EVs are trying to navigate those crosscurrents. Executives at Cadillac backtracked on a goal—boldly declared in the early-2020s EV frenzy—of having an all-electric lineup by 2030.[3] Caddy dealer Claude Burns doesn't yet regret spending all that money to prep his Rock Hill store to service EVs, but he's glad to have that new oil-change shop. Mercedes also came off its 2030-all-EV proclamation.[4]

But not one big carmaker has backed away. Factories are still going up. Models are still rolling out, backed by multimillion-dollar advertising campaigns. In the United States alone, about a hundred EV models are expected to be on the market by 2026. Not only do surveys show that about half of Americans are open to buying an EV, but younger people are far more inclined to give them a try than older buyers.[5]

What's also clear, though, is that gas–electric hybrid cars are going to figure more prominently in this transition than many carmakers had expected even just a couple years ago—a trend Akio Toyoda had foretold, even if he was attacked for that message. Most car executives—and investors—wanted to skip over that murky middle ground. The cost and complexity of tucking both an internal-combustion system *and* a battery-and-electric-motor setup under the sheet metal is a turnoff. So many execs figured, why would we do that?

They've gotten their answer: customers want them. Not enough people are ready to go fully electric without the perceived safety net of a gas engine onboard. That reckoning has sparked a bit of a hybrid frenzy. Automakers that had downplayed or ditched hybrids, like GM and VW, are left scrambling to reengage and come up with new ones. Companies with strong hybrid portfolios, like Toyota, Honda, and Ford, are leaning in. Ford said it would have hybrid versions of every vehicle in US showrooms by the end of the decade.

By 2024, Ford's Jim Farley was asking himself the $50 billion question every day: just how far will this EV story go? Will it be just a niche part of the market, or completely take over? Farley takes comfort in what he hears while talking to customers: *most* people like their EV more than the gas-burners they replaced. They're quiet, quick. People don't have to visit the gas station. But ultimately, he thinks a lot of people will also end up somewhere in the middle: turning to electrons to boost their driving experience, but maybe not a full EV yet. For so many years, the small, sensible Prius defined what a hybrid is, but that's changing to meet this middle consumer. Today, Ford's hybrid pickup truck can power your home for three days. Stellantis's Ram brand is jamming a big battery into a pickup truck that still has a 3.6-liter V6 gas engine to deliver nearly 700 miles of driving range.[6] Not to be outdone, Chinese electric-car powerhouse BYD in 2024 said it would have an electric-gas combo car that could go 1,250 miles without stopping to charge or refuel.[7]

Farley won't hazard a guess at where EV market share ends up and when. But he's confident that *electrified* cars—whether solely battery powered or supplemented by internal combustion—will dominate.

"The real story is that people will have way more options than they do today," Farley said. "They will be very rational about picking the one that makes sense for their lifestyle."[8] This echoes a TV spot for Ford that was in heavy rotation during sports events in the spring and summer of 2024, "More Choices," which makes virtue of the choice customers have between electric, hybrid, and gas.

Still, as all of this strategic murkiness plays out in the car business, one thing has become plainly clear: all eyes have turned to China. Anxiety about China's ability to take over the car business seems to be a strategic acknowledgment that the market will embrace EVs as the dominant mode of transport. Political and business leaders in these other countries feel pressure to slow that down, lest they lose the industry.

Over just a few months in the spring of 2024, both the United States and European Union slapped tariffs on Chinese imports of electric vehicles and batteries. Even though there wasn't an imminent threat to Chinese cars being sold on US soil, the Biden administration dropped the hammer with a 100 percent tariff. In the EV issue, two of Biden's top agenda items—climate change and job growth—were at odds. Yes, let's move to EVs as quickly as possible, as long as they're not made in China, so as to protect US jobs and all that investment going into places like BlueOval City, that massive factory complex in Haywood County, Tennessee, which is scheduled to open right about when this book publishes.

Even free-market Elon Musk was sounding alarm bells about Chinese EV makers. "If there are no trade barriers established, they will pretty much demolish most other car companies in the world," he said in 2024.[9] He later said that he doesn't support tariffs and doesn't think Tesla needs them to take on the Chinese carmakers.

In Europe, the threat wasn't theoretical, as those BYD cargo ships pulling into ports made viscerally clear. Chinese imports already accounted for 19 percent of the EVs sold in Europe by mid-2024—up from almost zero in 2019.[10] The European Union justified its tariffs with findings from a long-running investigation that determined Chinese government subsidies for its homegrown carmakers were creating an unlevel playing field.

Some car execs have adopted an "if you can't beat 'em join 'em" strategy, by partnering with Chinese automakers to develop more-affordable options. VW put $700 million into China's Xpeng, which will help the German automaker develop two VW-badged SUVs for the China market.[11] Stellantis invested $1.6 billion into China's Leapmotor.[12]

Many carmakers know, and have seen, that those consumers beyond the early adopters who do switch to EVs *love* them. They do not want to go back, even in a market where the infrastructure to support them, charging stations, and so forth, isn't fully built out yet. The challenge is creating the cars with the right mix of range and price and features—a mix the Chinese companies are way ahead on—to get more people to make the leap. Ultimately, carmakers are realizing that to navigate this messy transition, they need to appeal to more buyers like Jose Chao.

Chao doesn't fit the early-adopter mold that helped drive the early excitement around EVs. He is not overtly trying to save the planet with an EV purchase. Nor is he an Elon Musk disciple, or an engineer who's simply infatuated with the tech. An EV is not a status symbol for him. Chao is a sixty-two-year old accountant in Miami, who just happened to get interested in an EV "during all the media hype" of the past few years.

He had a "wow moment" after test-driving a Ford Mach-E. He later was lured by a massive price cut to trade in his Honda Civic to buy a Tesla Model Y, in 2023. He has some gripes, for sure. He admits to some range anxiety—especially on the highway, when he gets an uneasy feeling watching his miles click down faster than in city driving. There aren't enough public chargers around town. But he loves the car's quiet zip, the roomy cabin, and the frunk, where he stores his restaurant leftovers so that they don't stink up the interior.

"It's a nicer driving experience overall. It's different," Chao said. "I love it."[13]

NOTES

INTRODUCTION

1. Author interview, August 8, 2023.

2. International Energy Agency, "Cars and Vans," IEA website. https://www.iea.org/energy-system/transport/cars-and-vans.

3. Frédéric Simon, "Dan Lert: The City of Paris Aims to Phase Out Diesel by 2024, and Thermal Cars by 2030," Euractiv, June 24, 2022, https://www.euractiv.com/section/air-pollution/interview/dan-lert-the-city-of-paris-aims-to-phase-out-diesel-by-2024-and-thermal-cars-by-2030/.

4. "Tesla-the Coming of Age Electric Vehicles," *Inc. Magazine*. https://theincmagazine.com/tesla-the-coming-of-age-electric-vehicles/.

5. Joshua Davis, "How Elon Musk Turned Tesla Into the Car Company of the Future," *Wired*, September 27, 2010. https://www.wired.com/2010/09/ff-tesla/.

6. Tim Higgins, "Tesla Rivals GM as the Most Valuable Auto Maker in U.S.," *Wall Street Journal*, April 10, 2017, https://www.wsj.com/articles/tesla-overtakes-gm-to-become-most-valuable-u-s-auto-maker-1491832043.

7. "D11 Conference: Elon Musk Full Interview," YouTube, May 29, 2013. https://www.youtube.com/watch?v=UiPO4BUfov8.

8. Paul Lienert, "Exclusive: Automakers to Double Spending on EVs, Batteries to $1.2 Trillion by 2030," Reuters, October 25, 2022, https://www.reuters.com/technology/exclusive-automakers-double-spending-evs-batteries-12-trillion-by-2030-2022-10-21/.

9. Mike Colias, "Gas Engines, and the People Behind Them, Are Cast Aside for Electric Vehicles," Wall Street Journal, July 23, 2021, https://www.wsj.com/articles/gas-engines-cast-aside-electric-vehicles-job-losses-detroit-11627046285.

10. International Energy Agency, "Global EV Outlook 2024: Trends in Electric Cars," IEA website, https://www.iea.org/reports/global-ev-outlook-2024/trends-in-electric-cars.

11. Mike Wayland, "EV euphoria is dead. Automakers are Scaling Back or Delaying Their Electric Vehicle Plans," CNBC, March 13, 2024, https://www.cnbc.com/2024/03/13/ev-euphoria-is-dead-automakers-trumpet-consumer-choice-in-us.html.

12. Bloomberg Intelligence, "Despite Hurdles, Vehicle Electrification in the US Is Likely Here to Stay, Finds Bloomberg Intelligence," Bloomberg, April 4, 2024, https://www.bloomberg.com/company/press/despite-hurdles-vehicle-electrification-in-the-us-is-likely-here-to-stay-finds-bloomberg-intelligence/.

13. Dan Neil, "You've Formed Your Opinion on EVs. Now Let Me Change It.," *Wall Street Journal*, January 19, 2024, https://www.wsj.com/lifestyle/cars/youve-formed-your-opinion-on-evs-now-let-me-change-it-6c6fd1c1.

14. Ford fourth-quarter earnings call, February 6, 2024, https://s201.q4cdn.com/693218008/files/doc_financials/2023/q4/Ford-Q4-2023-Earnings-Call-Transcript.pdf.

15. International Energy Agency, "Global EV Outlook 2024: Trends in Electric Cars," IEA website, https://www.iea.org/reports/global-ev-outlook-2024/trends-in-electric-cars.

16. Ford fourth-quarter earnings call, February 6, 2024, https://s201.q4cdn.com/693218008 /files/doc_financials/2023/q4/Ford-Q4-2023-Earnings-Call-Transcript.pdf.

CHAPTER 1

1. "President Biden Drives Electric Hummer," C-Span via YouTube, November 17, 2021, https://www.youtube.com/watch?v=XrDGeAkWkAs.

2. Stephen Edelstein, "$12,500 EV Tax Credit, Union-Built Bonus Included in Plan Biden Claims Can Pass Congress," *Green Car Reports*, October 29, 2021, https://www.greencarreports .com/news/1134010_12-500-ev-tax-credit-union-built-bonus-biden-congress.

3. Author covered the event, November 17, 2021.

4. Nathan Bomey and Jaci Smith, "As GM Closes Additional Plants, Delaware Has Already Long Said Farewell," *USA Today*, November 28, 2018, https://www.delawareonline.com/story /money/business/2018/11/28/gm-general-motors-plant-closures-job-cuts/2127326002/.

5. Detroit Historical Society, "Encyclopedia of Detroit," https://detroithistorical.org/learn /encyclopedia-of-detroit/general-motors.

6. Paul B. MacCready, "Sunraycer Odyssey," Caltech Library, Engineering & Science, Winter 1988, https://calteches.library.caltech.edu/605/2/MacCready.pdf.

7. Brian Corey, "November 14, 1996–GM Releases Its EV1, First Electric Car," This Day in Automotive History, November 14, 2023, https://automotivehistory.org/gm-releases-ev1/.

8. *Los Angeles Times* staff, "Baywatch Actress Is 1 of 2 Arrested at Protest," *Los Angeles Times*, March 15, 2005, https://www.latimes.com/archives/la-xpm-2005-mar-15-me-briefs15.5-story.html.

9. Owen Edwards, "The Death of the EV-1," *Smithsonian Magazine*, June 2006, https://www .smithsonianmag.com/science-nature/the-death-of-the-ev-1-118595941/.

10. Author interview, November 2021.

11. Blake Z. Rong, "Stories With Bob Lutz," *AutoWeek*, November 25, 2013, https://www .autoweek.com/car-life/a1945191/stories-bob-lutz/.

12. David Welch, "Bob Lutz: The First Virtual Carmaker?," *Bloomberg BusinessWeek*, June 17, 2001, https://www.bloomberg.com/news/articles/2001-06-17/bob-lutz-the-first-virtual -carmaker?sref=PRBlrg7S.

13. Bob Lutz, *Car Guys vs. Bean Counters: The Battle for the Soul of American Business* (New York: Portfolio, 2011).

14. Kevin Krolicki, "GM Exec Stands by Calling Global Warming a 'Crock,'" Reuters, February 22, 2008, https://www.reuters.com/article/idUSN22372976/.

15. Lutz quotes in this chapter are from an author interview in November 2021 unless otherwise noted.

16. Frank Markus, "1966 GM Electrovan Fuel Cell Prototype Turns 50," *MotorTrend,* November 1, 2016, https://www.motortrend.com/news/1966-gm-electrovan-fuel-cell-prototype-turns-50/.

17. Top Gear team, "Chevrolet Volt Concept News - Detroit show: Chevrolet Volt Concept – 2007," BBC TopGear, Jan. 8, 2007, https://www.topgear.com/car-news/chevrolet-volt-concept-news -detroit-show-chevrolet-volt-concept-2007.

18. Carscoops, "Chevrolet Volt: Live Photos and Video Footage from the Detroit Presentation," Carscoops.com, September 16, 2008, https://www.carscoops.com/2008/09/chevrolet-volt-live -photos-and-video/.

19. Jim Motavalli, "G.M. Tones Down the Volt," *New York Times*, Sept. 21, 2008 https:// archive.nytimes.com/query.nytimes.com/gst/fullpage-9A0DEED9113DF932A1575AC0A96E9 C8B63.html.

20. Andrew Peterson, "2011 Chevrolet Volt Order Guide Shows Few Options, Pricing Due Tomorrow," MotorTrend, July 26, 2010, https://www.motortrend.com/news/2011-chevrolet-volt -order-guide-shows-few-options-pricing-due-tomorrow-8358/.

21. Phil LeBeau, "GM And Why It Believes The Volt Is A Winner," cnbc.com, September 16, 2008, https://www.cnbc.com/id/26739496.

22. "Chevrolet Volt: The Top Products of 2010," *Popular Mechanics*, September 27, 2010, https://www.popularmechanics.com/cars/hybrid-electric/a6141/chevrolet-volt-top-products-2010/.

23. Dan Neil, "Chevrolet Volt: A Win for the Home Team," October 29, 2010, https://www.wsj .com/articles/SB10001424052702304510704575562363168727230.

24. Chris Woodyard, "Obama's 'Secret': He Drove a Chevrolet Volt," *USA Today*, October 25, 2021.

25. George F. Will, "What's Driving Obama's Subsidies of Chevy's Volt?," *Washington Post*, November 14, 2010, https://www.washingtonpost.com/wp-dyn/content/article/2010/11/12 /AR2010111204494.html.

26. CBS News, "Yes, You Can Put A Gun Rack In A Chevrolet Volt," CBS News, February 24, 2012, https://www.cbsnews.com/texas/news/yes-you-can-put-a-gun-rack-in-a-chevrolet-volt/.

27. "Tesla Model S 2013-2014 Road Test," consumerreports.org video, https://www .consumerreports.org/video/view/cars/auto-test-track/2369062091001/tesla-model-s-2013-2014 -road-test/#:~:text=The%20high%2Dtech%20all%2Delectric,Consumer%20Reports%20has%20 ever%20tested.

28. Joann Muller, "Exclusive: Inside New CEO Mary Barra's Urgent Mission To Fix GM," Forbes, June 6, 2014, https://www.forbes.com/sites/joannmuller/2014/05/28/exclusive-inside-mary -barras-urgent-mission-to-fix-gm/.

29. Mike Colias, "The Incredible Shrinking GM: Mary Barra Bets That Smaller Is Better," Wall Street Journal, September 18, 2020, https://www.wsj.com/articles/the-incredible-shrinking-gm -mary-barra-finds-success-by-getting-smaller-11600438421.

30. Author covered the event, December 2013.

31. Bill Vlasic, "New G.M. Chief Is Company Woman, Born to It," New York Times, December 10, 2013, https://www.nytimes.com/2013/12/11/business/gm-names-first-female-chief-executive .html.

32. Mike Colias, "Mary Barra Spent a Decade Transforming GM. It Hasn't Been Enough," *Wall Street Journal*, December 16, 2023, https://www.wsj.com/business/autos/mary-barra-spent -a-decade-transforming-gm-it-hasnt-been-enough-d82f4c5a.

33. Mike Colias, "The Incredible Shrinking GM: Mary Barra Bets That Smaller Is Better," Wall Street Journal, September 18, 2020, https://www.wsj.com/articles/the-incredible-shrinking-gm -mary-barra-finds-success-by-getting-smaller-11600438421.

34. Author covered the event, October 2017.

35. Author covered the event, March 4, 2020.

36. Mike Colias, "The Incredible Shrinking GM: Mary Barra Bets That Smaller Is Better," *Wall Street Journal*, September 18, 2020, https://www.wsj.com/articles/the-incredible-shrinking -gm-mary-barra-finds-success-by-getting-smaller-11600438421.

CHAPTER 2

1. All Galyen quotes and recollections in this chapter are from author interviews, March 10, 2023 and June 5, 2023.

2. "Profile: Robin Zeng," Forbes, as of July 28, 2024, https://www.forbes.com/profile/robin-zeng/.

3. Mike Colias and Jennifer Hiller, "Proposed Tax Break for Buying Electric Vehicles Is Too Hard to Get, Auto Makers Say," *Wall Street Journal*, August 4, 2022, https://www.wsj.com/articles /auto-makers-ask-congress-for-easier-path-to-electric-vehicle-tax-break-11659614402.

4. "Responsible Sourcing," Cobalt Institute, https://www.cobaltinstitute.org/cobalt-sourcing-responsability/; "From Cobalt to Cars: How China Exploits Child and Forced Labor in the Congo," Congressional-Executive Commission on China, November 14, 2023, https://www.cecc.gov/events/hearings/from-cobalt-to-cars-how-china-exploits-child-and-forced-labor-in-the-congo#:~:text=80%25%20of%20the%20DRC's%20cobalt,battery%20makers%20around%20the%20world; Pete Pattisson and Febriana Firdaus, "Battery Arms Race': How China has Monopolised the Electric Vehicle Industry," the *Guardian*, November 25, 2021, https://www.theguardian.com/global-development/2021/nov/25/battery-arms-race-how-china-has-monopolised-the-electric-vehicle-industry.

5. "The Nobel Prize in Chemistry 2019," The Nobel Prize, 2019, https://www.nobelprize.org/prizes/chemistry/2019/popular-information/.

6. "The Nobel Prize in Chemistry 2019," The Nobel Prize, 2019, https://www.nobelprize.org/prizes/chemistry/2019/popular-information/.

7. Knvul Sheikh, Brian X. Chen, and Ivan Penn, "Lithium-Ion Batteries Work Earns Nobel Prize in Chemistry for 3 Scientists," *New York Times*, October 9, 2019, https://www.nytimes.com/2019/10/09/science/nobel-prize-chemistry.html.

8. "The Nobel Prize in Chemistry 2019," The Nobel Prize, 2019, https://www.nobelprize.org/prizes/chemistry/2019/popular-information/.

9. BYD, "Hello! We are BYD!," BYD website, December 29, 2022, https://www.byd.com/eu/blog/Hello-we-are-BYD.

10. Jack Goodman, "Has China Lifted 100 Million People Out of Poverty?," BBC.com, February 27, 2021, https://www.bbc.com/news/56213271.

11. "China's Economic Rise: History, Trends, Challenges, and Implications for the United States," Congressional Research Service, June 25, 2019, https://sgp.fas.org/crs/row/RL33534.pdf.

12. "Letter from Dr. Sun Yat-sen to Henry Ford and Ford's Response," *New York Times* archive, October 21, 2013, https://archive.nytimes.com/www.nytimes.com/interactive/2013/10/20/business/international/21ford-letters.html.

13. Zak Dychtwald, "China's New Innovation Advantage," *Harvard Business Review*, May–June 2021, https://hbr.org/2021/05/chinas-new-innovation-advantage.

14. "State Motor Vehicle Registrations, by Years 1900-1995," Federal Highway Administration, https://www.fhwa.dot.gov/ohim/summary95/mv200.pdf.

15. Unpublished interview by author, undisclosed person familiar with Ford's efforts, 2019.

16. David Welch, "Online Extra: Rick Wagoner on GM's Chinese Future," Bloomberg, June 20, 2004, https://www.bloomberg.com/news/articles/2004-06-20/online-extra-rick-wagoner-on-gms-chinese-future?sref=PRBlrg7S.

17. "China/Germany: VW China Sales Pass Home Market," Just Auto, July 27, 2009, https://www.just-auto.com/news/china-germany-vw-china-sales-pass-home-market/?cf-view.

18. "Detroit Looks East: GM Sells More Cars in China Than in the US," The Guardian, July 2, 2010, https://www.theguardian.com/business/andrew-clark-on-america/2010/jul/02/generalmotors-china.

19. Mike Floyd, "2007 Shanghai Auto Show Roundup: A Look Inside China's Latest Automotive Extravaganza," MotorTrend, April 24, 2007, https://www.motortrend.com/features/2007-shanghai-auto-show/.

20. Bloomberg staff, "The World's Leading Electric-Car Visionary Isn't Elon Musk," Bloomberg, September 26, 2018, https://www.bloomberg.com/news/features/2018-09-26/world-s-electric-car-visionary-isn-t-musk-it-s-china-s-wan-gang?sref=PRBlrg7S.

21. Uwe Parpart, "Green Energy Innovation to Fuel China's Tech Revolution," *Asia Times*, April 27, 2022, https://asiatimes.com/2022/04/green-energy-innovation-to-fuel-chinas-tech-revolution/.

22. Bloomberg staff, "Father of China's Electric-Car Industry Says His Friend Elon Musk Will Challenge Local Automakers," Bloomberg, June 13, 2019, https://www.bloomberg.com/news /features/2019-06-13/elon-musk-is-a-pretty-good-friend-wan-gang-says?sref=PRBlrg7S.

23. Keith Bradsher, "China Outlines Plans for Making Electric Cars," *New York Times*, April 10, 2009, https://www.nytimes.com/2009/04/11/business/energy-environment/11electric .html.

24. Bryan Walsh, "The World's Most Polluted Places," *Time*, September 13, 2007, https:// content.time.com/time/specials/2007/article/0,28804,1661031_1661028_1661017,00.html.

25. "Worldwatch Institute: 16 of World's 20 Most-Polluted Cities in China," Voice of America News, October 31, 2009 https://www.voanews.com/a/a-13-2006-06-28-voa36/397920.html.

26. Bob Davis, "Most of Beijing's Olympic Pollution Cleanup Evaporated a Year Later," *Wall Street Journal*, March 29, 2011, https://www.wsj.com/articles/BL-REB-13630.

27. CBC News staff, "China's No Car Day Has Little Impact on Driving Habits," CBC News, September 22, 2007, https://www.cbc.ca/news/world/china-s-no-car-day-has-little-impact-on -driving-habits-1.650358#:~:text=Facebook-,-,Motorists%20appeared%20to%20largely%20 ignore%20China's%20first%20No%20Car%20Day,in%20droves%3A%20ride%20a%20bicycle.

28. Stephen Edelstein, "China to Replace 70,000 Gasoline Cabs With Electric Cars," Green Car Reports, March 7, 2017, https://www.greencarreports.com/news/1109224_china-to-replace-70000 -gasoline-cabs-with-electric-cars.

29. Trefor Moss, "China, With Methodical Discipline, Conjures a Market for Electric Cars," Wall Street Journal, Oct. 2, 2017, https://www.wsj.com/articles/china-with-methodical-discipline -takes-global-lead-in-electric-cars-1506954248.

30. Trefor Moss, "China, With Methodical Discipline, Conjures a Market for Electric Cars," *Wall Street Journal*, October 2, 2017, https://www.wsj.com/articles/china-with-methodical -discipline-takes-global-lead-in-electric-cars-1506954248.

31. "Tesla's Musk Laughs at BYD," YouTube, November 15, 2011, https://www.youtube.com /watch?v=_9ftbRWqkj0.

32. River Davis and Selina Cheng, "How China's BYD Became Tesla's Biggest Threat," *Wall Street Journal*, October 4, 2023, https://www.wsj.com/business/autos/how-chinas-byd-became -teslas-biggest-threat-5edfd080.

33. Milken Institute, executive biography, Global Conference 2024, https://milkeninstitute.org /events/a0c1u00000cf196uab/speakers/stella-li.

34. Author interviews, former BYD employees, June 2023.

35. Davis and Cheng, "How China's BYD Became Tesla's Biggest Threat."

36. Matt Hardigree, "Detroit Auto Show: World Exclusive Surreal, Illegal Test Drive Of Chinese Hybrid Through Cobo Arena," Jalopnik, Jan. 15, 2008, https://jalopnik.com/detroit-auto -show-world-exclusive-surreal-illegal-tes-344806

37. "This Chinese City Has More Than 16,000 Electric Buses," World Economic Forum/Quartz, January 5, 2018, https://www.weforum.org/agenda/2018/01/this-chinese-city-has-more-than -16-000-electric-buses/.

38. Alex Crippen, "Fortune Puts Warren Buffett In 'Car of the Future' Driver's Seat," cnbc. com, August 5, 2010, https://www.cnbc.com/2009/04/13/fortune-puts-warren-buffett-in-car-of -the-future-drivers-seat.html#:~:text=Munger%20describes%20Wang%20to%20Fortune,never% 20seen%20anything%20like%20it.%E2%80%9D.

39. All Cheng quotes and reflections in this chapter are from an author interview on June 29, 2023.

40. Alisa Priddle, "Ram, Maserati Levante Launches Disappoint Outspoken FCA CEO," *MotorTrend*, Aprile 26, 2018, https://www.motortrend.com/news/ram-maserati-launches -disappoint-outspoken-fca-ceo/.

41. NIO, "The Story of NIO's Design Headquarters," *NIO blog*, September 13, 2017, https://www.nio.com/blog/story-nios-design-headquarters?&noredirect.

42. TopGear team, "Nio ES6 review," July 24, 2019. https://www.topgear.com/car-reviews/nio/es6.

CHAPTER 3

1. All quotes and reflections from Yeung in this chapter are from author interviews in February and March, 2023; and February 2024.

2. Phoebe Wall Howard, "CEO Jim Farley Vows Ford Motor Will Not Split in Two—Previews Restructuring," *Detroit Free Press*, February 23, 2022.

3. "EV Powertrain Components, Basics," EVReporter.com, October 28, 2019, https://evreporter.com/ev-powertrain-components/.

4. Csaba Csere, "2008 Tesla Roadster Road Test," March 1, 2008, https://www.caranddriver.com/reviews/a15150030/2008-tesla-roadster-road-test/.

5. All quotes and reflections from Penkevich in this chapter are from author interviews in May 2021 and February 2023.

6. Mike Colias, "Gas Engines, and the People Behind Them, Are Cast Aside for Electric Vehicles," *Wall Street Journal*, July 23, 2021, https://www.wsj.com/articles/gas-engines-cast-aside-electric-vehicles-job-losses-detroit-11627046285.

7. "Critical Supplier Strategies: Five Key Forces Amidst a Plateauing Market," S&P Global Mobility (formerly IHS Markit), March 7, 2019.

8. "1964 Ford Mustang Hardtop 260 V-8 (man. 3) Detailed Performance Review, Speed vs rpm and Accelerations Chart," automobile catalogue, https://www.automobile-catalog.com/performance/1964/829115/ford_mustang_hardtop_260_v-8.html#gsc.tab=0.

9. Murilee Martin, "Class of 1965: When GM Had Eight V8 Engine Families," The Truth About Cars, December 18, 2010, https://www.thetruthaboutcars.com/2010/12/class-of-1965-when-gm-had-eight-v8-engine-families/.

10. Reginald Stuart, "G.M. Calls Its Engine Swapping Innocent, But to the Brand-Faithful Buyer It's a Sin," *New York Times*, March 15, 1977, https://www.nytimes.com/1977/03/15/archives/article-4-no-title-engine-swaps-innocent-to-gm-but-sinful-to.html?auth=login-google1tap&login=google1tap

11. Tom Halter, "Automotive History: The 1977 Oldsmobile Chevrolet Engine Scandal – There's No Rocket In My 88's Pocket," *Curbside Classic*, May 13, 2020, https://www.curbsideclassic.com/automotive-histories/automotive-history-the-1977-oldsmobile-chevrolet-engine-scandal/.

12. "The EPA 2023 Automotive Trends Report," December 2023, https://www.epa.gov/system/files/documents/2023-12/420s23002.pdf.

13. Eric A. Taub, "Start-Stop Technology Is Spreading (Like It or Not)," *New York Times*, April 7, 2016, https://www.nytimes.com/2016/04/08/automobiles/wheels/start-stop-technology-is-coming-to-cars-like-it-or-not.html.

14. Elisabeth Behrmann, Birgit Jennen, and Christoph Rauwald, "Daimler Gets Slapped With Recall, But Avoids Risk of Fines," Bloomberg, June 12, 2019, https://www.bloomberg.com/news/articles/2018-06-11/germany-orders-daimler-to-recall-774-000-diesel-cars-in-europe?sref=PRBlrg7S.

CHAPTER 4

1. "Volkswagen Type 2, 1950–2025," classic.com, https://www.classic.com/m/volkswagen/type-2/.

2. All quotes and reflections from Montoya in this chapter are from author interviews, February 2024.

3. Eric Jaffe, "The Study That Brought Down Volkswagen," Bloomberg, September 24, 2015, https://www.bloomberg.com/news/articles/2015-09-24/the-west-virginia-study-that-started-the-volkswagen-scandal?sref=PRBlrg7S.

4. Nick Gibbs, "Gasoline Engines Poised to Gain Share in Europe," *Automotive News Europe*, November 22, 2012, https://europe.autonews.com/article/20121122/ANE/311229999/gasoline-engines-poised-to-gain-share-in-europe.

5. Volkswagen, "VW Beetle, The Real Miracle," company website, https://www.volkswagen-newsroom.com/en/the-volkswagen-beetle-a-success-story-2341/vw-beetle-the-real-miracle-2356.

6. Sam Abuelsamid, "Volkswagen kicks off cross-country DIESELUTION tour," *Autoblog*, September 26, 2007, https://www.autoblog.com/2007/09/26/volkswagen-kicks-off-cross-country-dieselution-tour/.

7. Eric Jaffe, "The Study That Brought Down Volkswagen," Bloomberg, September 24, 2015, https://www.bloomberg.com/news/articles/2015-09-24/the-west-virginia-study-that-started-the-volkswagen-scandal?sref=PRBlrg7S.

8. "Eight More VW Employees Charged in Diesel Scandal," *Automotive News Europe*, September 23, 2020, https://europe.autonews.com/automakers/eight-more-vw-employees-charged-diesel-scandal.

9. "Volkswagen Says Diesel Scandal has Cost it 31.3 Billion Euros," Reuters, March 17, 2020, https://www.reuters.com/article/idUSKBN2141JA/.

10. Edward Taylor and Jan Schwartz, "Bet Everything on Electric: Inside Volkswagen's Radical Strategy Shift," Reuters, February 6, 2019, https://www.reuters.com/article/idUSKCN1PV0K0/.

11. Andreas Cremer and Jan Schwartz, "Volkswagen Accelerates Push Into Electric Cars with $40 Billion Spending Plan," Reuters, November 17, 2017, https://www.reuters.com/article/idUSKBN1DH13R/.

12. Author interview, February 2024.

13. "2017 Detroit Auto Show—Volkswagen Press Conference," YouTube, August 20, 2017, https://www.youtube.com/watch?v=pcrOe1VuqhQ.

14. Andrew J. Hawkins, "Why Volkswagen Keeps Making Microbus Throwbacks It Never Intends to Sell," *The Verge*, Jan. 10, 2017, https://www.theverge.com/2017/1/10/14215476/volkswagen-microbus-concept-id-buzz-emissions-scandal.

15. *Handelsblatt* staff, "Change Management; VW Boss Wants to Make History as Well as Cars," *Handelsblatt*, November 20, 2018.

16. Christoph Rauwald and Francine Lacqua, "Volkswagen CEO Confident He Can Catch Tesla in E-Car Race," Bloomberg, January 24, 2020, https://www.bloomberg.com/news/articles/2020-01-24/volkswagen-ceo-confident-he-can-catch-tesla-in-electric-car-race?sref=PRBlrg7S.

17. Christoph Rauwald, "Tesla Is No Niche Automaker Anymore, Volkswagen's CEO Says," Bloomberg, October 24, 2019, https://www.bloomberg.com/news/articles/2019-10-24/volkswagen-s-ceo-says-tesla-is-no-niche-automaker-anymore?sref=PRBlrg7S.

18. Edward Taylor, "Tesla CEO Met VW CEO During Germany Visit: Source," Reuters via Yahoo!, September 4, 2020, https://finance.yahoo.com/news/tesla-ceo-met-vw-ceo-151557195.html.

19. Fred Lambert, "VW Admits Tesla's Lead in Software and Self-Driving in Internal Leak," Electrek, April 27, 2020, https://electrek.co/2020/04/27/vw-admits-tesla-lead-software-leak-internal/.

20. "Diess to VW Workers: 'I'm Worried About Wolfsburg,'" Reuters, via *Automotive News Europe*, November 4, 2021, https://europe.autonews.com/automakers/diess-vw-workers-im-worried-about-wolfsburg.

21. Melissa Eddy, "He's Steering Volkswagen, With His Eyes on Beating Tesla," *New York Times*, November 29, 2021, https://www.nytimes.com/2021/11/24/business/volkswagen-ceo-herbert-diess.html.

22. "Diess to VW Workers: 'I'm Worried about Wolfsburg,'" Reuters, via *Automotive News Europe*, November 4, 2021, https://europe.autonews.com/automakers/diess-vw-workers-im -worried-about-wolfsburg.

CHAPTER 5

1. All Jankowsky quotes and reflections in this chapter are from author interviews in December 2021 and July 2023.

2. Alliance for Automotive Innovation, "Get Connected Electric Vehicle Quarterly Report," second quarter 2023, https://www.autosinnovate.org/posts/papers-reports/Get%20 Connected%20EV%20Quarterly%20Report%202023%20Q2.pdf?utm_source=Sailthru&utm _medium=Newsletter&utm_campaign=Auto-File&utm_term=092723.

3. Young People of the Next Generation, "Renewable Energy Company Founder Mentor – David Jankowsky," https://ypng.co/mentors/6132/.

4. Beth Wallis, "The road to electric: Oklahoma navigates transition to embracing electric vehicles," StateImpact Oklahoma, March 3, 2022, https://stateimpact.npr.org/oklahoma/2022/03/03 /the-road-to-electric-oklahoma-navigates-transition-to-embracing-electric-vehicles/.

5. Author interview with David Jankowsky, July 2023.

6. IEA, "Trends in Electric Vehicle Charging," Global EV Outlook 2024, https://www.iea.org /reports/global-ev-outlook-2024/trends-in-electric-vehicle-charging.

7. Tajammul Pangarkar, "Electric Vehicle Charging Infrastructure Statistics: New Technology," scoop.market.us, February 19, 2024, https://scoop.market.us/electric-vehicle -charging-infrastructure-statistics/.

8. Selina Xu, "This Chinese Province Has More EV Chargers Than All of the US," Bloomberg, October 21, 2022, https://www.bloomberg.com/news/articles/2022-10-22/this-chinese -province-has-three-times-more-ev-chargers-than-all-of-the-us?sref=PRBlrg7S.

9. Selina Xu, "This Chinese Province Has More EV Chargers Than All of the US."

10. J.D. Power, "Action Needed to Keep Charging from Short Circuiting EV Purchase Consideration, J.D. Power Finds," press release, June 15, 2023, https://www.jdpower.com/business /press-releases/2023-us-electric-vehicle-consideration-evc-study.

11. Gabe Shenhar and Alex Knizek, "Can Electric Vehicle Owners Rely on DC Fast Charging?," *Consumer Reports*, November 7, 2022.

12. Isabella Sullivan, "Cheaper and Cleaner: Electric Vehicle Owners Save Thousands," *NRDC Expert* (blog), March 11, 2024, https://www.nrdc.org/bio/isabella-sullivan/cheaper-and -cleaner-electric-vehicle-owners-save-thousands#:~:text=Depending%20on%20the%20model%20 analyzed,time%2Dof%2Duse%20rates.

13. Dustin Hawley, "How Much Does It Cost To Install An EV Charger?," J.D. Power Car Shopping Guides, December 11, 2022, https://www.jdpower.com/cars/shopping-guides/how -much-does-it-cost-to-install-an-ev-charger.

14. US Census Bureau, "Census Bureau Estimates Show Average One-Way Travel Time to Work Rises to All-Time High," March 18, 2021, https://www.census.gov/newsroom/press-releases/2021 /one-way-travel-time-to-work-rises.html.

15. IEA, "Global EV Outlook 2023: Trends in Charging Infrastructure," https://www.iea.org /reports/global-ev-outlook-2023/trends-in-charging-infrastructure.

16. "EV Leasing Volumes Poised to Surge as Tax Rule Makes It Cheaper to Lease than Buy," J.D. Power's E-Vision Intelligence Report, May 2023, https://www.jdpower.com/business /resources/ev-leasing-volumes-poised-surge-tax-rule-makes-it-cheaper-lease-buy.

17. John Voelcker, "GM Won't Fund CCS Fast-Charging Sites for 2017 Chevy Bolt EV," *Green Car Reports*, January 13, 2016, https://www.greencarreports.com/news/1101774_gm-wont-fund-ccs-fast-charging-sites-for-2017-chevy-bolt-ev.

18. Tesla, "Tesla Motors Launches Revolutionary Supercharger Enabling Convenient Long Distance Driving," press release, September 24, 2012.

19. David R. Baker, "Why Tesla's EV Charging Plugs Are Becoming Industry Standard," Bloomberg, July 7, 2023, https://www.bloomberg.com/news/articles/2023-07-07/why-tesla-ev-charging-plugs-are-becoming-industry-standard?sref=PRBlrg7S.

20. Author interview, August 8, 2023.

21. "Ford and Tesla Supercharger Announcement – Full Jim Farley and Elon Musk Twitter Spaces," YouTube, Farzad Mesbahi, May 25, 2023, https://www.youtube.com/watch?v=rsrs4a212g0.

22. Mike Colias, River Davis, and Ryan Felton, "Big Automakers Plan Thousands of EV Chargers in $1 Billion U.S. Push," *Wall Street Journal*, July 26, 2023, https://www.wsj.com/articles/big-automakers-plan-thousands-of-ev-chargers-in-1-billion-u-s-push-af748d19.

23. Author interview, July 6, 2023.

24. U.S. Drive, "Summary Report on EVs at Scale and the U.S. Electric Power System," U.S. Drive, November 2019, https://www.energy.gov/eere/vehicles/articles/summary-report-evs-scale-and-us-electric-power-system-2019.

25. Chris Harto, "Blog: Can the Grid Handle EVs? Yes!" *Consumer Reports*, May 10, 2023, https://advocacy.consumerreports.org/research/blog-can-the-grid-handle-evs-yes/.

26. KPMG, "Place Your Billion-Dollar Bets Wisely," 2022, https://assets.kpmg.com/content/dam/kpmg/xx/pdf/2021/08/place-your-billion-dollar-bets-wisely.pdf.

27. All Li quotes and reflections in this chapter are from author interviews, September 2023 and March 2024.

CHAPTER 6

1. Author interviews, 2021.

2. Author interview with Darren Palmer, January 3, 2022.

3. General Motors Keynote- CES 2022, YouTube, January 6, 2022, https://www.youtube.com/watch?v=rDAwAeSVljY.

4. Joseph Spak, "Quick Auto Takes from CES: GM Presentation Underwhelming? RIVN Sell-Off Overdone?" RBC Capital Markets research note, Jan. 5, 2022.

5. Author interview with a person at the meetings, January 2022.

6. "Number of Tesla Cars Sold in Norway from 2009 to 2022," via Statista, 2023, https://www.statista.com/statistics/419267/tesla-car-sales-in-norway/.

7. Author interview, May 10, 2022.

8. Bill Vlasic, "A Star at Toyota, a Believer at Ford," *New York Times*, April 20, 2008, https://www.nytimes.com/2008/04/20/business/20ford.html.

9. Author interview, January 14, 2020.

10. Author interview, January 14, 2020.

11. Gerald Donaldson, "Drivers / Hall of Fame: Phil Hill," Formula1.com, https://www.formula1.com/en/drivers/hall-of-fame/Phil_Hill.html.

12. "1955 Best of Show Winner," Pebble Beach Concours, April 4, 2020, https://pebblebeachconcours.net/history-traditions/1955-best-of-show-winner/.

13. Bill Vlasic, *Once Upon a Car*, 2011, William Morrow, pg. 5.

14. Author interview, January 14, 2020.

15. Author interviews, 2021–2023.

16. Cars.com, "Ford Focus Electric," n.d., https://www.cars.com/research/ford-focus
_electric/.

17. Author interview, January 15, 2018.

18. Michael Martinez, "Bill Ford Didn't Want to Call Mach-E a Mustang—Until He Drove It,"
Automotive News Europe, November 19, 2019, https://europe.autonews.com/automakers
/bill-ford-didnt-want-call-mach-e-mustang-until-he-drove-it.

19. Author interview, May 10, 2022.

20. Author Interview, January 24, 2023.

21. Charles Morris, "Proterra Beefs Up Its Battery Expertise: Q&A with CTO Dustin Grace,"
Charged Electric Vehicles Magazine, November 11, 2020, https://chargedevs.com/features
/proterra-beefs-up-its-battery-expertise-qa-cto-dustin-grace/.

22. Author interview, January 24, 2023.

23. Author interview, May 10, 2022.

24. Author interview with Darren Palmer, January 3, 2022.

25. Author interview with Jim Hackett, April 7, 2022.

26. Michael Wayland, "Forget 'Ford v Ferrari,' the Mustang Mach-E takes a shot at Tesla,"
CNBC.com, November 18, 2019, https://www.cnbc.com/2019/11/18/forget-ford-v-ferrari-ford
-targets-tesla-with-mustang-mach-e.html.

27. Dan Woodland "How Luther Heartthrob Idris Elba Went From Building Ford Fiestas in
Dagenham with His Dad to Superhero-Stardom," *Daily Mail*, January 8, 2024, https://www.dailymail
.co.uk/news/article-12937929/Idris-Elba-Hackney-council-estate-Wire-knife-crime.html.

28. Author covered the event, November 2019.

29. Paul Lienert, "Ford Caps F-150 Lightning Orders at 200,000—CEO Farley," Reuters,
December 9, 2021, https://www.reuters.com/business/autos-transportation/ford-caps-f-150
-lightning-orders-200000-ceo-farley-2021-12-09/.

CHAPTER 7

1. Matthew Johnson, "Rivian IPO: What Happened and Why it Matters," Investopedia,
November 15, 2021, https://www.investopedia.com/rivian-ipo-what-happened-and-why-it
-matters-5209505.

2. Anita Hamilton, "How Tesla Rejoined the Trillion Dollar Club," *Barron's*, March 30, 2022,
https://www.barrons.com/visual-stories/tesla-trillion-dollar-value-stock-split-01648654777.

3. Author interviews, 2022–2023.

4. Ben Foldy, "The Challenging Road Ahead for Rivian's Billionaire CEO," *Wall Street
Journal*, December 11, 2021, https://www.wsj.com/articles/the-challenging-road-ahead-for-rivians
-billionaire-ceo-11639198848.

5. Steven Cole Smith, "Before His Battery Behemoths, Rivian's Billionaire Founder Made an
Eco Sports Car," Hagerty, November 21, 2022, https://www.hagerty.com/media/automotive-history
/before-his-battery-behemoths-rivians-billionaire-founder-made-an-eco-sports-car/.

6. Kate Birch, "Meet the CEO: RJ Scaringe of Electric Vehicle Maker Rivian," *Business Chief*,
November 3, 2021, https://businesschief.com/technology-and-ai/meet-ceo-rj-scaringe-electric
-vehicle-maker-rivian.

7. Ben Foldy, "The Challenging Road Ahead for Rivian's Billionaire CEO," *Wall Street Journal*,
Dec. 11, 2021, https://www.wsj.com/articles/the-challenging-road-ahead-for-rivians-billionaire-ceo
-11639198848.

8. Ben Foldy, "The Challenging Road Ahead for Rivian's Billionaire CEO."

9. Wayne T. Price, "Local Firm to Develop New Hybrid Car," *Florida Today*, May 22, 2010.

10. Ben Foldy, "The Challenging Road Ahead for Rivian's Billionaire CEO."

11. Wayne T. Price, "Local Firm to Develop New Hybrid Car."

12. Ryan Denham, "Meet the Auto Industry Veterans Who Guided Rivian Through 'Pivot Points,'" *WGLT Sound Ideas*, November 30, 2018, https://www.wglt.org/show/wglts-sound-ideas/2018-11-30 /meet-the-auto-industry-veterans-who-guided-rivian-through-pivot-points.

13. Abdul Latif Jameel IPR Company Limited, "Our Story," company website, https://alj.com /en/about/story/.

14. Ben Foldy, "The Challenging Road Ahead for Rivian's Billionaire CEO."

15. Matthew Martin, "Rivian's IPO Mints $11.5 Billion Fortune for Saudi Investor," Bloomberg, November 10, 2021, https://www.bloomberg.com/news/articles/2021-11-10/rivian-s -ipo-delivers-near-9-billion-fortune-for-saudi-investor.

16. Akin Oyedele, "A 1960s 'Chicken War' Reveals the Driving Force Behind the Boom of American Pickup Trucks," *Business Insider*, June 22, 2018, https://www.businessinsider.com /american-trucks-chicken-tax-explains-domination-2018-6.

17. "Rivian Automotive to Build R&D Center for High-Tech Vehicles in Mich," *Automotive News*, November 24, 2015, https://www.autonews.com/article/20151124/OEM06/151129924 /rivian-automotive-to-build-r-d-center-for-high-tech-vehicles-in-mich.

18. Mitsubishi Motors, "Mitsubishi Motors Achieves Five Million Vehicle Sales in the United States," *Mitsubishi News*, December 28, 2015, https://media.mitsubishicars.com/en-US/releases /mitsubishi-motors-achieves-five-million-vehicle-sales-in-the-united-states.

19. Author interview, March 3, 2023.

20. Brian Solomon, "Is Uber Trying To Kill Lyft With A Price War?," *Forbes*, January 25, 2016, https://www.forbes.com/sites/briansolomon/2016/01/25/is-uber-trying-to-kill-lyft-with-a-price -war/?sh=761bab056573.

21. Author interview, March 3, 2023.

22. Sarah Nardi, "Is Coffee Hound the Brew that Launched a Thousand Rivian R1Ts?," *WGLT Local News*, February 22, 2022. https://www.wglt.org/local-news/2022-02-22/is-coffee-hound -the-brew-that-launched-a-thousand-rivian-r1ts.

23. Bernie Woodall and Paul Lienert, "Rivian Bids on Shuttered Mitsubishi Plant in Illinois," Reuters, December 9, 2016, https://www.reuters.com/article/us-autos-rivian-mitsubishimotors /rivian-bids-on-shuttered-mitsubishi-plant-in-illinois-idUSKBN13Y1SZ/.

24. Harry Brumpton and Stephen Nellis, "Amazon, GM in Talks to Invest in Electric Pickup Truck Maker Rivian: Sources," Reuters, February 13, 2019, https://www.reuters.com/article/us -rivian-electric-amazon-com-gm-exclusi/amazon-gm-in-talks-to-invest-in-electric-pickup-truck -maker-rivian-sources-idUSKCN1Q12PV/.

25. Author interview, October 2021.

26. Author interview, February 21, 2023.

27. Author interview, October 2021.

28. Rivian, "Rivian Launches World's First Electric Adventure Vehicles," Rivian newsroom, November 27, 2018, https://rivian.com/newsroom/article/rivian-launches-worlds-first-electric -adventure-vehicles.

29. Chris Hamilton, "Faux-150: Rivian Test Mules Disguised as Ford F-150s," *Street Trucks*, March 8, 2019, https://www.streettrucksmag.com/rivian-f150-disguise/.

30. Guest Contributor, "Electric Vehicle Sales in US Hit the Accelerator Pedal—Even Beyond California," *Clean Technica*, September 23, 2023, https://cleantechnica.com/2023/09/22/electric -vehicle-sales-in-us-hit-the-accelerator-pedal/#:~:text=More%20widespread%20adoption%20 of%20EVs,a%20more%20conservative%2Dleaning%20population.

31. Rivian, AutoMobility LA, Los Angeles Auto Show, press announcement, 2018, https:// www.youtube.com/watch?v=y0CPRUgvoVo.

32. "Best Cars of the 2018 Los Angeles Auto Show: MotorTrend Favorites," *MotorTrend*, Nov. 30, 2018, https://www.motortrend.com/features/best-cars-of-the-2018-los-angeles-auto-show -motortrend-favorites/.

33. Neal E. Boudette, "Amazon Invests in Rivian, a Tesla Rival in Electric Vehicles," *New York Times*, February 15, 2019, https://www.nytimes.com/2019/02/15/business/rivian-amazon.html.

34. Michael Matthews, "Chevy's CMO Responds to Ford's Cease and Desist of 'Mayan' Super Bowl Spot," *Forbes*, February 6, 2012, https://www.forbes.com/sites/michaelmatthews/2012/02/06/chevys-cmo-responds-to-fords-cease-and-desist-of-mayan-super-bowl-spot/?sh=c3cfb5353649.

35. Melissa Burden, "Chevy Targets Ford F-150 Aluminum Beds in New Ads," *Detroit News*, June 8, 2016, https://www.detroitnews.com/story/business/autos/general-motors/2016/06/08/chevy-targets-ford-aluminum-beds-ads/85589760/.

36. Author interview, October 2021.

37. David Welch, Spencer Soper, and Keith Naughton, "GM, Amazon Deal May Give Electric Truck Maker Rivian a Spark," Bloomberg, Feb. 12, 2019, https://www.bloomberg.com/news/articles/2019-02-13/gm-amazon-backing-would-give-plug-in-truck-maker-rivian-a-spark?sref=PRBlrg7S.

38. Author interview, October 2021.

39. Todd Lassa, "Ford Invests $500 Million in Rivian, Opens New Path to F-150 EV," *MotorTrend*, April 24, 2019, https://www.motortrend.com/news/ford-invests-500-million-rivian-ev-f-150/.

CHAPTER 8

1. Elle Kaia, "Waiting Times for New Electric Car Deliveries Down by 39% Since October Peak,"*Electrifying.com* (blog), July 7, 2023, https://www.electrifying.com/blog/article/waiting-times-for-new-electric-car-deliveries-down-by-42-since-october-peak.

2. iea, "Trends in Electric Light-duty Vehicles," iea website, n.d., https://www.iea.org/reports/global-ev-outlook-2023/trends-in-electric-light-duty-vehicles.

3. Joe Lorio, "2024 Toyota bZ4X," *Car and Driver*, 2024, https://www.caranddriver.com/toyota/bz4x.

4. Hollis quotes and reflections in this chapter are from an author interviews on December 1, 2022.

5. Noah Baustin, "Which EV Cars Are Most Popular? See the Tata by County," *San Francisco Standard*, May 12, 2023, https://sfstandard.com/2023/05/12/california-electric-car-market-share-zero-emission-vehicles-tesla/.

6. Brendan Mcaleer, "How the Toyota Prius Brought Hybrids to the Mainstream," *Car and Driver*, December 14, 2022, https://www.caranddriver.com/features/a42230942/toyota-prius-hybrid-history/.

7. Micheline Maynard, "Say 'Hybrid' and Many People Will Hear 'Prius,'" *New York Times*, July 4, 2007, https://www.nytimes.com/2007/07/04/business/04hybrid.html; Washington Post Staff Writer, "Half Gas, Half Electric, Total California Cool," *Washington Post*, June 5, 2002, https://www.washingtonpost.com/archive/lifestyle/2002/06/06/half-gas-half-electric-total-california-cool/5f0a18f2-e6af-44b8-a3d8-eb750fa4b462/.

8. Toyota, "Toyota Sells One-Millionth Prius in the U.S.," Toyota Newsroom, April 6, 2011, https://pressroom.toyota.com/toyota-sells-one-millionth-prius-in-the-u-s/.

9. Fred Lambert, "Toyota's 'Self-Charging Hybrid' Ad Is Nanned in Norway, Deemed a Lie," Electrek, January 24, 2020, https://electrek.co/2020/01/24/toyota-self-charging-hybrid-ad-banned-norway-lie/.

10. Eric C. Evarts, "Commentary: Toyota Corolla Hybrid Ad Brags About Not Plugging In," Green Car Reports, February 15, 2019, https://www.greencarreports.com/news/1121517_commentary-toyota-corolla-hybrid-ad-brags-about-not-plugging-in.

11. Matt Wolfe, "Toyoda to Toyota," Automotive Hall of Fame, September 12, 2016, https://www.automotivehalloffame.org/toyoda-to-toyota/.

12. Patrick George, "Toyota Chairman Akio Toyoda Doesn't Plan to Quit Racing Anytime Soon, *Road and Track*, June 22, 2023, https://www.roadandtrack.com/news/a44299507/toyota-chairman-akio-toyoda-doesnt-plan-to-quit-racing-anytime-soon/.

13. Toyota, "President Akio Toyoda's Speech at CES 2018," Toyota press release, January 9, 2018, https://global.toyota/en/newsroom/corporate/20566886.html.

14. Piers Ward, "Why the Toyota Hilux Matters in Africa," *Autocar*, December 7, 2020, https://www.autocar.co.uk/opinion/advice/why-toyota-hilux-matters-africa.

15. Colin Velez, "Toyota EVP Jack Hollis on EVs, Vehicle Affordability, and Inventory," *CBT News*, March 6, 2024, https://www.cbtnews.com/toyota-evp-jack-hollis-on-the-evs-vehicle-affordability-and-inventory/.

16. Ada Kong, Mingyang Zheng, Xixi Zhang, Greenpeace: "Auto Environmental Guide 2021: A Comparative Analysis," pp 3, 5, 73, https://www.greenpeace.org/static/planet4-eastasia-stateless/2021/11/47de8bb4-gpea_auto_environmental_guide_2021.pdf.

17. Richart Truett, "Margo Oge's Faves: Maria Callas, Odysseus and Lots of Shoes," *Automotive News*, October 1, 2007, https://www.autonews.com/article/20071001/SUB/710010337/margo-oge-s-faves-maria-callas-odysseus-and-lots-of-shoes.

18. Oge quotes and reflections in this chapter are from an author interview on December 19, 2023.

19. Author interview, December 19, 2023.

20. Office of the Press Secretary, "Obama Administration Finalizes Historic 54.5 MPG Fuel Efficiency Standards," White House, August 28, 2012, https://obamawhitehouse.archives.gov/the-press-office/2012/08/28/obama-administration-finalizes-historic-545-mpg-fuel-efficiency-standard.

21. Bill Vlasic, "Carmakers Back Strict New Rules for Gas Mileage," *New York Times*, February 7, 2013, https://www.nytimes.com/2011/07/29/business/carmakers-back-strict-new-rules-for-gas-mileage.html%20s.

22. Austin Weber, "Assembling Ford's Aluminum Wonder Truck," *Assembly*, March 3, 2015, https://www.assemblymag.com/articles/92728-assembling-fords-aluminum-wonder-truck.

23. Transport Policy, "California: ZEV," TransportPolicy.net, https://www.transportpolicy.net/standard/california-zev/#:~:text=In%20January%202012%2C%20CARB%20formally,new%20vehicle%20sales%20by%202025.

24. Steve Greenfield, "How Tesla Is Banking Billions In Regulatory Emissions Credits," *CBT News*, February 16, 2024, https://www.cbtnews.com/how-tesla-is-banking-billions-in-regulatory-emissions-credits/#:~:text=That%20brought%20the%20cumulative%20total,be%20both%20cool%20and%20profitable.

25. Author interviews of people with knowledge of the matter.

26. Ron Cogan, "First Toyota RAV4 EV for Consumers," *Green Car Journal*, September 7, 2022, https://greencarjournal.com/dont-miss/first-toyota-rav4-ev-for-consumers/.

27. Author interview, Dec. 20, 2023.

28. David Shepardson, "Trump Administration Moves Closer to Rolling Back U.S. Vehicle Fuel Economy Increases," Reuters, January 14, 2020, https://www.reuters.com/article/us-autos-emissions/trump-administration-moves-closer-to-rolling-back-u-s-vehicle-fuel-economy-increases-idUSKBN1ZE011/.

29. Mike Colias, Ben Foldy, and Andrew Restuccia, "The Auto Industry Wanted Easier Environmental Rules. It Got Chaos," *Wall Street Journal*, February 3, 2020, https://www.wsj.com/articles/the-auto-industry-wanted-easier-environmental-rules-it-got-chaos-11580745826.

30. Mike Colias, Ben Foldy, and Andrew Restuccia, "The Auto Industry Wanted Easier Environmental Rules. It Got Chaos."

31. "The Automotive Industry's Earnest Views on Carbon Neutrality-JAMA Press Conference," *Toyota Times*, September 13, 2021, https://toyotatimes.jp/en/toyota_news/169.html.

32. Hans Greimel, "Akio Toyoda: All-EV plans Are A Threat To Japan," *Automotive News*, September 19, 2021, https://www.autonews.com/mobility-report/toyota-president-akio-toyoda -all-ev-plan-wrong-japan.

33. Peter Landers, "Toyota's Chief Says Electric Vehicles Are Overhyped," *Wall Street Journal*, December 17, 2020, https://www.wsj.com/articles/toyotas-chief-says-electric-vehicles-are -overhyped-11608196665.

34. East Peterson-Trujillo, "EVs or Obsolescence. Which will Toyota choose?," *Public Citizen*, November 7, 2022, https://www.citizen.org/news/evs-or-obsolescence-toyota-choose/.

35. David Dolan, "Toyota Heads Into AGM Under Pressure from Pension Funds Over Climate," Reuters, June 14, 2022, https://www.reuters.com/business/autos-transportation/toyota -heads-into-agm-under-pressure-pension-funds-over-climate-2022-06-14/.

36. Hans Greimel, "As Green Reputation Takes a Beating, Toyota Fights Back," *Automotive News*, November 22, 2021, https://www.autonews.com/mobility-report/toyota-fights.

37. Norihiko Shirouzu, Joseph White, and Maki Shiraki, "Focus: Toyota Looks to Overhaul EV Strategy as New CEO Takes Charge," Reuters, April 6, 2023, https://www.reuters.com/business /autos-transportation/toyota-looks-overhaul-ev-strategy-new-ceo-takes-charge-2023-04-06./

38. Mark Rechtin, "From An Odd Couple To A Dream Team," *Automotive News*, August 13, 2012, https://www.autonews.com/article/20120813/OEM03/308139960/from-an-odd-couple-to-a -dream-team.

39. Norihiko Shirouzu, "Exclusive: Toyota Scrambles for EV Reboot with Eye on Tesla," Reuters, October 24, 2022, https://www.reuters.com/business/autos-transportation/exclusive -toyota-scrambles-ev-reboot-with-eye-tesla-2022-10-24/.

40. Norihiko Shirouzu, Joseph White, and Maki Shiraki, "Focus: Toyota Looks to Overhaul EV Strategy as New CEO Takes Charge," Reuters, April 6, 2023, https://www.reuters.com/business /autos-transportation/toyota-looks-overhaul-ev-strategy-new-ceo-takes-charge -2023-04-06./

41. River Davis, "Toyota Rethinks EV Strategy With New CEO," *Wall Street Journal*, January 29, 2023, https://www.wsj.com/articles/toyota-akio-toyoda-koji-sato-evs-electric-vehicles-new -ceo-11675008222.

42. Davis, "Toyota Rethinks EV Strategy With New CEO."

CHAPTER 9

1. John Huetter, "GM: Cadillac to be 'Tip of the Corporate Spear' on Electric Vehicle," *Repair Driven News*, January 14, 2019, https://www.repairerdrivennews.com/2019/01/14/gm-cadillac-to -be-tip-of-the-corporate-spear-on-electric-vehicles/.

2. Matt Posky, "Auto Dealers Report 2021 Profits Will Break Previous Record," *The Truth About Cars*, January 3, 2022, https://www.thetruthaboutcars.com/2022/01/auto-dealers-report-2021 -profits-will-break-previous-record/.

3. Mike Colias, "About 150 U.S. Cadillac Dealers to Exit Brand, Rather Than Sell Electric Cars," *Wall Street Journal*, December 4, 2020, https://www.wsj.com/articles/about-150-u-s -cadillac-dealers-to-exit-brand-rather-than-sell-electric-cars-11607111494.

4. Diego Rosenberg, "Cadillac Dealers Feeling The Pressure From de Nysschen," *GM Authority* (blog), October 20, 2014, https://gmauthority.com/blog/2014/10/cadillac-dealers-feeling-the-pressure -from-de-nysschen/.

5. Burns's quotes and reflections in this chapter are from author interviews in July 2021 and May 2023.

6. Peter Valdes-Depena, "Why Car Shopping Is So Bizarre in the United States," CNN Business, May 9, 2022, https://www.cnn.com/2022/04/30/cars/why-we-buy-cars-this-way/index.html.

7. Beepi Inc., "Study: Americans Feel Taken Advantage of at the Car Dealership," PR Newswire, July 21, 2016, https://www.prnewswire.com/news-releases/study-americans-feel-taken-advantage-of-at-the-car-dealership-300301866.html.

8. Nick Zamanov, "The Impact of Electric Vehicles on Car Dealerships," Cyber Switching, July 11, 2023, https://cyberswitching.com/impact-electric-vehicles-car-dealerships/#:~:text=Unlike%20traditional%20combustion%20engine%20vehicles,generated%20from%20service%20and%20repairs.

9. Benjamin Preston, "Pay Less for Vehicle Maintenance With an EV," *Consumer Reports*, September 26, 2020, https://www.consumerreports.org/car-repair-maintenance/pay-less-for-vehicle-maintenance-with-an-ev/.

10. Nick Gibbs, "Agency Sales Model Faces Big Test," *Automotive News Europe*, June 22, 2023, https://europe.autonews.com/retail/automakers-agency-direct-sales-retail-model-faces-big-test.

11. Nora Eckert and Mike Colias, "Ford's Entrenched Problems Complicate CEO Jim Farley's EV Future," *Wall Street Journal*, February 16, 2023, https://www.wsj.com/articles/ford-ceo-jim-farley-grapples-with-problems-from-the-auto-makers-past-b098271c; "Reuters: Ford Will Challenge Dealers to Match Tesla's Lower Selling Costs," Nada, September 9, 2022, https://www.nada.org/nada/nada-headlines/reuters-ford-will-challenge-dealers-match-teslas-lower-selling-costs.

12. "Ford CEO Farley Says Electric Vehicles Will Be Sold 100% Online, Have Nonnegotiable Price," *Detroit Free Press*, June 1, 2022, https://www.freep.com/story/money/cars/ford/2022/06/01/ford-electric-vehicles-online-sales-non-negotiable-price/7468899001/.

13. Author interview, May 15, 2023.

14. Author interview, May 15, 2023.

CHAPTER 10

1. "Tesla Gross Margin 2010-2024 | TSLA," *Macrotrends*, 2024, https://www.macrotrends.net/stocks/charts/TSLA/tesla/gross-margin#:~:text=2021%2D12%2D31,2017%2D03%2D31.

2. Emailed statement to author from Hau Thai-Tang, August 2024.

3. Farley quotes in this chapter are from an August 8, 2023 interview unless otherwise noted.

4. Nimisha Jain, "EVs Top September's Car Sales in Europe Thanks to Tesla," *AutoTrader*, October 26, 2021, https://www.autotrader.co.uk/content/news/tesla-model-3-tops-september-car-sales-in-europe?refresh=true.

5. Bloomberg, "How Elon Musk Built a Tesla Factory in China in Less Than a Year," *Fortune*, January 7, 2020, https://fortune.com/2020/01/07/elon-musk-tesla-gigafactory-shanghai-china-ceremony/.

6. Simon Alvarez, "Tesla Model Y Sets Records By Being Europe's Best-Selling Car for 2023," Teslarati, January 19, 2024, https://www.teslarati.com/tesla-model-y-europe-best-selling-car-2023-records/#:~:text=1%2C%20and%20it%20will%20also,selling%20cars%20list%20for%202023.

7. ET Spotlight Special, "Tesla's Giga Berlin Will Ease the Surge in Demand for EVs in Europe," *Economic Times*, March 22, 2022, https://economictimes.indiatimes.com/news/international/us/teslas-giga-berlin-will-ease-the-surge-in-demand-for-evs-in-europe/articleshow/90381124.cms.

8. Anita Hamilton, "How Tesla Rejoined the Trillion Dollar Club," *Barrons*, March 29, 2022, https://www.barrons.com/visual-stories/tesla-trillion-dollar-value-stock-split-01648654777.

9. Rebecca Elliot, "Tesla Posts Record $3.3 Billion Quarterly Profit," *Wall Street Journal*, April 20, 2022, https://www.wsj.com/articles/tesla-earnings-q1-2022-11650413167.

10. Ford, proxy statement, March 31, 2023, https://corporate.ford.com/content/dam/corporate/us/en-us/documents/reports/notice-of-the-2023-virtual-annual-meeting-of-shareholders-and-proxy-statement.pdf.

11. Michael Martinez, "How Ford Plans to Woo Top Tech Talent," *Automotive News*, March 2, 2022, https://www.autonews.com/executives/how-ford-wooing-top-tech-talent-new-ev-unit.

12. Author interviews with people who were present, May 2023.

13. Lora Kolodny, "Elon Musk says Tesla was 'About a Month' from Bankruptcy During Model 3 Ramp," CNBC, November 3, 2020, https://www.cnbc.com/2020/11/03/musk-tesla-was-about -a-month-from-bankruptcy-during-model-3-ramp.html#:~:text=In%20a%20Twitter%20 conversation%20on,%2D2017%20to%20mid%2D2019.

14. Helena Vieira, "With Software Updates, Tesla Upends Product Life Cycle in the Car Industry," *LSE* (blog), January 31, 2017, https://blogs.lse.ac.uk/businessreview/2017/01/31/with -software-updates-tesla-upends-product-lifecycle-in-the-car-industry/.

15. Mike Colias, "Why Your Car Will Become Even More Like an iPhone," *Wall Street Journal*, November 4, 2021, https://www.wsj.com/articles/why-your-car-will-become-even-more-like-an -iphone-11636038092.

16. Author interview, October 18, 2021.

17. Ford, "Ford+ Delivers Solid 2023, Provides Outlook for Healthy '24; Company Declares Regular, Supplemental Stock Dividends," 4th quarter earnings press release, February 6, 2024.

18. Nora Eckertt and Rebecca Elliot, "Ford, Tesla CEOs Exchange Jabs and Praise Amid Heated EV Rivalry," *Wall Street Journal,* May 7, 2023, https://www.wsj.com/articles/ford-tesla -ceos-exchange-jabs-and-praise-amid-heated-ev-rivalry-6526555d.

19. Scott Evans, "2024 Tesla Model 3 vs. 2023 Tesla Model Y: Y U Should Wait," *MotorTrend*, November 10, 2023, https://www.motortrend.com/reviews/2024-tesla-model-3-vs-2023-tesla -model-y-comparison-test-review/.

20. Steve Blank, "Apple's Marketing Playbook Was Written in the 1920s," *Atlantic*, October 26, 2011, https://www.theatlantic.com/business/archive/2011/10/apples-marketing -playbook-was-written-in-the-1920s/247417/.

21. Author interview, October 18, 2021.

22. Ford, "Ford Hybrids, EVS, Transit Set Records; Q1 Sales Top Industry, Up 7%," press release, April 2, 2024, https://media.ford.com/content/fordmedia/fna/us/en/news/2024/04/03 /ford-hybrids--evs--transit-set-records--q1-sales-top-industry--u.html.

23. Ford, "Ford Previews Effect on Parts Shortages on Q3 Performance, Reaffirms Full-Year Adjusted EBIT Guidance of $11.5B-$12.5B" press release, September 19, 2022, https://media.ford .com/content/fordmedia/fna/us/en/news/2022/09/19/ford-previews-effect-of-parts-shortages-on -q3-performance.html.

24. Ford, fourth-quarter 2022 earnings conference call, February 2, 2023, https://s201.q4cdn .com/693218008/files/doc_financials/2022/q4/q4-22-earnings-transcript-2.2.23-final-(updated -feb.-15-2023).pdf.

25. Neal E. Boudette, "Ford Halts Production of Electric Pickup over Battery Issue," *New York Times*, February 14, 2023, https://www.nytimes.com/2023/02/14/business/ford-f150-lightning -battery.html.

26. Phoebe Wall Howard, "Ford Cuts Price of 2023 Mustang Mach-E by up to $8,100, Offers 0% Financing," *Detroit Free Press*, February 20, 2024, https://www.freep.com/story/money/cars /ford/2024/02/20/ford-mustang-mach-e-price-lower/72661435007/.

27. Former Contributor, "Tesla Slashes Prices Up To 20 Percent, Sending Shockwaves Through EV Industry," *Forbes*, January 26, 2023, https://www.forbes.com/sites/qai/2023/01/26/tesla-slashes -prices-up-to-20-percent-sending-shockwaves-through-ev-industry/?sh=307282793a28.

28. David Shepardson, "Ford Cuts Prices of Electric Mustang Mach-E by up to $8,100," Reuters, February 20, 2024, https://www.reuters.com/business/autos-transportation/ford-cuts -prices-all-variants-mustang-mach-e-electric-suv-2024-02-20/#:~:text=The%20No.,price%20 by%20%248%2C100%20to%20%2448%2C895.

29. Caleb Miller, "2023 Ford F-150 Lightning Base Price Drops by Roughly $10,000," *Car and Driver*, July 17, 2023, https://www.caranddriver.com/news/a44564936/2023-ford-f-150-lightning -price-cuts/.

30. Motley Fool, "Ford Motor Company (F) Q2 2023 Earnings Call Transcript," *Motley Fool*, July 27, 2023, https://www.fool.com/earnings/call-transcripts/2023/07/28/ford-motor-company -f-q2-2023-earnings-call-transcr/.

31. Author covered the trip including interview, August 8, 2023.

CHAPTER 11

1. Mark Olade, "Well-Known Salton Sea Origin Story Questioned by New Research, Suggesting it Wasn't 'Accidental,'" *Desert Sun*, May 8, 2020, https://www.desertsun.com/story /news/environment/2020/05/08/well-known-salton-sea-origin-story-questioned-new-research /3083262001/.

2. Associated Press, "History of the Salton Sea," *Marin Independent Journal*, June 2, 2015, https://www.marinij.com/2015/06/02/history-of-the-salton-sea/.

3. Lee Thomas-Mason, "Lost and Abandoned in the Desert: Salton Sea and Salvation Mountain," *Far Out*, October 9, 2017, https://faroutmagazine.co.uk/lost-and-abandoned-in-the -desert-salton-sea-and-salvation-mountain/.

4. Sonny Bono Salton Sea National Wildlife Refuge, "About Us," US Fish & Wildlife Service, https://www.fws.gov/refuge/sonny-bono-salton-sea/about-us.

5. Eric Lindberg, "As Salton Sea Shrinks, Experts Fear Far-Reaching Health Consequences," *USC Today*, August 28, 2019, https://today.usc.edu/salton-sea-shrinking-asthma-respiratory-health -air-quality/.

6. Dave Goodman, Patrick Mirick, and Kyle Wilson, "Salton Sea Geothermal Development," Pacific Northwest National Laboratory, June 2022, https://www.pnnl.gov/main/publications /external/technical_reports/PNNL-32717.pdf.

7. Carlo Cariaga, "Lawrence Berkeley National Lab to Lead Research on Salton Sea Geothermal Lithium," *Think Geoenergy*, February 21, 2022, https://www.thinkgeoenergy.com /berkeley-lab-to-lead-research-on-salton-sea-geothermal-lithium/.

8. Carolyn Gramling, "The Search for New Geologic Sources of Lithium Could Power a Clean Future," *ScienceNews*, May 7, 2019, https://www.sciencenews.org/article/search-new-geologic -sources-lithium-could-power-clean-future.

9. Teague Egan, "What is the Difference Between Hard Rock vs. Brine Lithium Sources?,"*Energyx*, April 25, 2023, https://energyx.com/blog/what-is-the-difference-between -hard-rock-vs-brine-lithium-sources/.

10. Muhammad Jamal Akbar, "Lithium Reserves by Country: Top 15 Countries," Insider Monkey via Yahoo! Finance, June 24, 2023, https://finance.yahoo.com/news/lithium-reserves -country-top-15-184450553.html.

11. Ara Persson, "Understand Lithium Mining's Environmental Impact," *Carbon Chain* (blog), March 8, 2024, https://www.carbonchain.com/blog/understand-lithium-minings-environmental -impact#:~:text=Carbon%20emissions%20of%20lithium%20produced,lithium%20produced%20 through%20brine%20extraction.

12. Maeve Campbell, "In Pictures: South America's 'Lithium Fields' Reveal the Dark Side of Our Electric Future," *Euronews*, January 2, 2022, https://www.euronews.com/green/2022/02/01 /south-america-s-lithium-fields-reveal-the-dark-side-of-our-electric-future.

13. Martin Silva, "South America's 'Lithium Triangle' Communities Are Being 'Sacrificed' to Save the Planet," *Euronews*, October 28, 2022, https://www.euronews.com/green/2022/10/28 /south-americas-lithium-triangle-communities-are-being-sacrificed-to-save-the-planet.

14. Joshua Longmore, "California Governor: 'We Are the Saudi Arabia of Lithium,'" *National*, February 23, 2022, https://www.thenationalnews.com/world/us-news/2022/02/23/california-governor-we-are-the-saudi-arabia-of-lithium/.

15. June Kim, "As Companies Eye Massive Lithium Deposits in California's Salton Sea, Locals Anticipate a Mixed Bag," *Inside Climate News*, August 26, 2023, https://insideclimatenews.org/news/26082023/salton-sea-lithium-mining-california/; California Energy Commission, "Selective Recovery of Lithium from Geothermal Brines," March 2020, https://www.energy.ca.gov/sites/default/files/2021-05/CEC-500-2020-020.pdf

16. Ines Ferré, "Crashing Lithium Prices Turn the Industry from 'Euphoria' to 'Despair.' What's Next?," Yahoo Finance, February 20, 2024, https://finance.yahoo.com/news/crashing-lithium-prices-turn-the-industry-from-euphoria-to-despair-whats-next-184543769.html?guccounter=1.

17. Nathan Gomes, "Global Demand for Lithium Batteries to Leap Five-Fold by 2030- Li-Bridge," Reuters, February 15, 2023; Jeff St. John, "The US EV industry Now Faces a Choice: Tax Credits or Chinese Batteries," Canary Media, December 6, 2023, https://www.canarymedia.com/articles/clean-energy-supply-chain/the-us-ev-industry-now-faces-a-choice-tax-credits-or-chinese-batteries.

18. Bill Whitaker, "Companies Develop Lithium Extraction for Batteries in California as U.S. Auto Industry Goes Electric," CBS News, May 7, 2023, https://www.cbsnews.com/news/lithium-extraction-california-electric-vehicle-batteries-60-minutes-transcript-2023-05-07/.

19. Author interview, August 15, 2023.

20. Jeff Clemetson, "ESM Poised to Power Lithium's 'White Gold Rush,'" *San Diego Business Journal*, October 3, 2022, https://www.sdbj.com/featured/esm-poised-to-power-lithiums-white-gold-rush/.

21. Mike Colias and Scott Patterson, "The New EV Gold-Rush: Automakers Scramble to Get Into Mining," *Wall Street Journal*, May 15, 2023, https://www.wsj.com/articles/the-new-ev-gold-rush-automakers-scramble-to-get-into-mining-ebda14eb.

22. The Benchmark Source, "Lithium Industry Needs Over $116 Billion to Meet Automaker and Policy Targets by 2030," *Benchmark Source*, August 4, 2023, https://source.benchmarkminerals.com/article/lithium-industry-needs-over-116-billion-to-meet-automaker-and-policy-targets-by-2030.

23. Timothy Noah and Adam Behsudi, "Trump Faces Failing Strategy on Auto Jobs as He Heads to Ohio," *PoliticoPro*, March 19, 2019, https://subscriber.politicopro.com/article/2019/03/trump-faces-failing-strategy-on-auto-jobs-as-he-heads-to-ohio-1282884.

24. Johnathan Lopez, "GM-Posco Plant Construction Under Way In Quebec," *GM Authority* (blog), February 10, 2023, https://gmauthority.com/blog/2023/02/gm-posco-plant-construction-under-way-in-quebec/#:~:text=A%20new%20cathode%20production%20plant,in%20future%20GM%20electric%20vehicles.

25. Reuters, "VW's PowerCo, Stellantis, Glencore Back $1B Nickel, Copper SPAC Deal in Brazil,", *Automotive News Europe*, June 12, 2023, https://europe.autonews.com/suppliers/vws-powerco-stellantis-gencore-back-1b-mine-deal-brazil.

26. Hallgerton & Company, "The Mining Supercycle and Its Demise," *FM Global*, December 4, 2015, https://www.mining.com/web/the-mining-supercycle-and-its-demise/.

27. Benchmark Source, "More Than 300 New Mines Required to Meet Battery Demand by 2035," *Benchmark Source*, September 6, 2022, https://source.benchmarkminerals.com/article/more-than-300-new-mines-required-to-meet-battery-demand-by-2035.

28. Julian Busch, "LG Chem Invests in Battery Joint Venture with GM," *Korea Certification*, October 2020, https://www.korea-certification.com/en/lg-chem-invests-in-battery-joint-venture-with-gm//.

29. Mike Colias and Scott Patterson, "The New EV Gold-Rush: Automakers Scramble to Get Into Mining," Wall Street Journal, May 15, 2023. https://www.wsj.com/articles/the-new-ev-gold-rush-automakers-scramble-to-get-into-mining-ebda14eb.

30. Author interview, March 3, 2023.

31. Talon, "Tesla and Talon Metals Enter Into Supply Agreement for Nickel," Talon Metals Corp., January 10, 2022, https://talonmetals.com/tesla-and-talon-metals-enter-into-supply-agreement-for-nickel/#:~:text=Agreement%20Highlights%3A,from%20the%20Tamarack%20Nickel%20Project.

32. Smruthi Nadig, "Top Ten Nickel-Producing Countries in 2023," Mining Technology, March 25, 2024, https://www.mining-technology.com/features/top-ten-nickel-producing-countries-in-2023/.

33. Sierra Club, "Critical Minerals: Nickel," Sierra Club North Star Chapter, March 26, 2024, https://www.sierraclub.org/minnesota/blog/2024/03/critical-minerals-nickel#:~:text=The%20proposed%20Tamarack%20nickel%20mine,and%20exploration%20for%20twenty%20years.

34. Author interview, March 2023.

35. Mike Colias and Scott Patterson, "The New EV Gold-Rush: Automakers Scramble to Get Into Mining," Wall Street Journal, May 15, 2023. https://www.wsj.com/articles/the-new-ev-gold-rush-automakers-scramble-to-get-into-mining-ebda14eb.

36. Mike Colias and Scott Patterson, "The New EV Gold-Rush: Automakers Scramble to Get Into Mining," Wall Street Journal, May 15, 2023, https://www.wsj.com/articles/the-new-ev-gold-rush-automakers-scramble-to-get-into-mining-ebda14eb; Ernest Scheyder, "UPDATE 4-GM to Help Lithium Americas Develop Nevada's Thacker Pass Mine," Yahoo Finance, January 31, 2023, https://finance.yahoo.com/news/1-gm-lithium-americas-develop-115126948.html.

37. Ernest Scheyder, "Insight: Inside the Race to Remake Lithium Extraction for EV Batteries," Reuters, June 16, 2023, https://www.reuters.com/markets/commodities/inside-race-remake-lithium-extraction-ev-batteries-2023-06-16/.

38. Wyatt Myskow, "Mining Companies Say They Have a Better Way to Get Underground Lithium, but Skepticism Remains," Inside Climate News, March 24, 2024, https://insideclimatenews.org/news/24032024/direct-lithium-extraction-green-river-utah/.

39. Goldman Sachs, "Direct Lithium Extraction: A Potential Game Changing Technology," Goldman Sachs, April 27, 2023, https://www.goldmansachs.com/intelligence/pages/gs-research/direct-lithium-extraction/report.pdf.

40. Sammy Roth, "The Salton Sea Has Even More Lithium Than Previously Thought, New Report Finds," Los Angeles Times, November 28, 2023, https://www.latimes.com/environment/newsletter/2023-11-28/the-salton-sea-has-even-more-lithium-than-previously-thought-new-report-finds-boiling-point.

41. Author visit, August 7, 2023.

42. Daniel Propp, "Can Direct Lithium Extraction Scale?," Latitude Media, April 11, 2024, https://www.latitudemedia.com/news/can-direct-lithium-extraction-scale.

43. Thomas Fudge, "Report Details Rich Lithium Deposits in Imperial County," KPBS, December 1, 2023, https://www.kpbs.org/news/science-technology/2023/12/01/report-details-rich-lithium-deposits-in-imperial-county.

44. Staff Report, "EnergySource Minerals Announces Contract with Ford for Geothermal Lithium," Imperial Valley Press, May 28, 2023, https://www.ivpressonline.com/featured/energysource-minerals-announces-contract-with-ford-for-geothermal-lithium/article_e9d6ccf6-fb4a-11ed-98bf-933253bc1638.html.

CHAPTER 12

1. Carlton Veirs, "The Men Who Built the Megasite," hellohaywood.com, https://hellohaywood .com/the-men-who-built-the-megasite/.

2. US census data, Haywood County, Tennessee, QuickFacts, https://www.census.gov /quickfacts/fact/table/haywoodcountytennessee/INC910222.

3. Julian McTizic, "'Destination Hatchie'-A System of Parks on The Historic Hatchie River," University of Tennessee Center for Industrial Services, August 15, 2019.

4. US census data, Haywood County, Tennessee.

5. Ashli Blow, "Ford Megasite Atop 'Recharge Zone' for Underregulated Memphis Sands Aquifer," *Tennessee Lookout*, January 3, 2022, https://tennesseelookout.com/2022/01/03/ford -megasite-atop-recharge-zone-for-underregulated-memphis-sands-aquifer/.

6. The Great Migration (1910–1970), National Archives, https://www.archives.gov/research /african-americans/migrations/great-migration.

7. Joyce Shaw Peterson, "Black Automobile Workers in Detroit, 1910–1930," *Journal of Negro History*, Summer 1979, https://www.jstor.org/stable/2717031.

8. Christopher L. Foote, Warren C. Whatley, and Gavin Wright, "Arbitraging a Discriminatory Labor Market: Black Workers at the Ford Motor Company, 1918–1947," *Journal of Labor Economics* 21, no. 3 (July 2003): 493–532, https://www.jstor.org/stable/10.1086/374957.

9. Thomas J. Sugrue, "From Motor City to Motor Metropolis: How the Automobile Industry Reshaped Urban America," Automobile in American Life and Society, http://www.autolife.umd .umich.edu/Race/R_Overview/R_Overview.htm.

10. Nora Eckert, "EV Boom Remakes Rural Towns in the American South," *Wall Street Journal*, August 31, 2023, https://www.wsj.com/business/autos/ev-plants-southern-states -ford-blueoval-city-2783da97.

11. Echo Day, "Black Farmers Talk About Collaboration, Paradigm Shift Amid Blue Oval City Eminent Domain Worries," *Leader*, June 29, 2023, https://covingtonleader.com/news/black -farmers-talk-about-collaboration-paradigm-shift-amid-blue-oval-city-eminent-domain-worries/.

12. eia (US Energy Information Administration);

13. Nora Eckert and Mike Colias, "The UAW-Ford Deal: What's in the Contract, Who Won and What It Means for GM and Stellantis," *Wall Street Journal*, October 26, 2023, https://www.wsj .com/business/autos/uaw-ford-labor-contract-strike-452197d5.

14. Carlton Veirs, "The Men Who Built the Megasite," *Hello Haywood!*, https://hellohaywood .com/the-men-who-built-the-megasite/.

15. All quotes and reflections from Banks in this chapter are from author interviews, August 28, 2023; September 21, 2023.

16. News Staff, "State Building Commission Approves West Tennessee Megasite Project," Clarksvilleonline.com, September 30, 2009.

17. Chas Sisk, "Tennessee's Auto Plant Ambitions Could Harm Hatchie River," *Tennessean*, May 18, 2014, https://www.tennessean.com/story/news/environment/2014/05/17/tennessees-auto -plant-ambitions-harm-h15atchie-river/9190349/.

18. Erik Schelzig, "Wastewater Line Approved for Megasite," *Jackson Sun*, August 22, 2016, https://www.jacksonsun.com/story/money/business/2016/08/22/megasite-use-35-mile-wastewater -line-mississippi-river/89108336/.

19. Author interviews, August 28, 2023; September 21, 2023.

20. Tyler Whetstone, "Randy Boyd Runs, Literally, Through East Tennessee on Gubernatorial Campaign," *Knox News*, August 22, 2017, https://www.knoxnews.com/story/news/politics /elections/2017/08/22/randy-boyd-runs-literally-through-east-tennessee-gubernatorial -campaign/570050001/.

21. Nora Eckert, "Ford's U.S. Auto Sales Fell 2% in 2022," *Wall Street Journal*, January 5, 2023, https://www.wsj.com/articles/fords-u-s-auto-sales-fell-2-in-2022-11672929951.

22. Rudy Williams, "Stanton, Tennessee, Residents Excited for Ford's BlueOval City," WATN-TV 24 Memphis, September 26, 2022.

23. Lea & Simmons Funeral Group obituary, "Franklin Smith May 21, 1950 — April 24, 2023," https://www.leaandsimmonsfuneralhome.com/obituaries/franklin-smith.

CHAPTER 13

1. All Grub quotes and reflections in this chapter are from an author interview and site visit on August 17, 2023.

2. Ford, "Ford to Lead America's Shift to Electric Vehicles With New BlueOval City Mega Campus in Tennessee and Twin Battery Plants in Kentucky," press release, September 27, 2021, https://corporate.ford.com/articles/electrification/blue-oval-city.html#:~:text=Ford's%20%247%20billion%20investment%20is,create%20a%20sustainable%20American%20manufacturing.

3. Heejin Kim, "Hyundai Raises 2030 Electric-Vehicle Sales Goal to 2 Million," Bloomberg, June 20, 2023, https://www.bloomberg.com/news/articles/2023-06-20/hyundai-raises-electric-vehicle-sales-goal-to-2-million-by-2030?sref=PRBlrg7S.

4. Christoph Steitz, "VW Shows Confidence in Electric Future with Higher Margin Goal," Reuters, July 13, 2021, https://www.reuters.com/business/autos-transportation/volkswagen-aims-half-vehicles-sales-be-electric-by-2030-2021-07-13/.

5. Mike Colias, "GM to Phase Out Gas- and Diesel-Powered Vehicles by 2035," *Wall Street Journal*, January 28, 2021, https://www.wsj.com/articles/gm-sets-2035-target-to-phase-out-gas-and-diesel-powered-vehicles-globally-11611850343.

6. Phoebe Sedgman, Jinshan Hong, and Linda Lew, "China's Stranglehold on EV Supply Chain Will Be Tough to Break," Bloomberg, September 27, 2023.

7. Paul Lienert, "Exclusive: Automakers to Double Spending on EVs, Batteries to $1.2 Trillion by 2030," Reuters, October 25, 2022, https://www.reuters.com/technology/exclusive-automakers-double-spending-evs-batteries-12-trillion-by-2030-2022-10-21/.

8. "GDP Ranked by Country 2024," World Population Review, https://worldpopulationreview.com/countries/by-gdp.

9. Center For Automotive Research, emailed research to author, June 7, 2024.

10. Rick Newman, "A Factory Boom is Finally Happening," Yahoo! Finance, June 15, 2023, https://finance.yahoo.com/news/a-factory-boom-is-finally-happening-190048801.html.

11. John Irwin, "The Microchip Shortage is Fading, but the Changes It Wrought Remain," *Automotive News*, December 22, 2023, https://www.autonews.com/manufacturing/microchip-shortages-impact-auto-industry-persists.

12. JP Morgan Research, "Inflation and the Auto Industry: When Will Car Prices Drop?," February 22, 2023, https://www.jpmorgan.com/insights/global-research/autos/when-will-car-prices-drop.

13. Megan Ruggles, "GM Secures 3 More Supply Agreements to Boost EV Production," Supply Chain Dive, August 4, 2022, https://www.supplychaindive.com/news/gm-secures-more-supply-agreements-for-electric-vehicles/628515/.

14. Nishtha Gupta, "Li-ion Cell Manufacturing: A Look at Processes and Equipment," *Emerging Technology News*, June 10, 2021, https://etn.news/energy-storage/li-ion-cell-manufacturing.

15. Cotes (industry supplier), "What is a Battery Dry Room, and Why Does the Air Inside Need Extremely Low Dewpoints?," https://www.cotes.com/faq/what-is-a-battery-dry-room.

16. Rebecca Bellan, "Tracking the EV Battery Factory Construction Boom Across North America," TechCrunch, August 16, 2023, https://techcrunch.com/2023/08/16/tracking-the-ev -battery-factory-construction-boom-across-north-america/.

17. All Kiefer quotes and reflections in this chapter are from an author interview, August 30, 2022.

18. Cameron McWhirter and Mike Colias, "Why Ford Picked Tennessee for Its New Electric-Vehicle Plant," *Wall Street Journal*, October 15, 2021, https://www.wsj.com/articles/why-ford -picked-tennessee-for-its-new-electric-vehicle-plant-11634302800.

19. Author site visit, August 17, 2023.

20. Keith Laing and Justin Sink, "Biden Flanked by Detroit Automakers Signs Clean Car Targets," Bloomberg, August 5, 2021, https://www.bloomberg.com/news/articles/2021-08-05 /biden-flanked-by-detroit-automakers-calls-for-clean-car-future?sref=PRBlrg7S.

21. White House, "Biden-Harris Administration Announces New Standards and Major Progress for a Made-in-America National Network of Electric Vehicle Chargers," press release, February 15, 2023, https://www.whitehouse.gov/briefing-room/statements-releases/2023/02/15 /fact-sheet-biden-harris-administration-announces-new-standards-and-major-progress-for-a -made-in-america-national-network-of-electric-vehicle-chargers/#:~:text=Today%2C%20the%20 Biden%2DHarris%20Administration,road%20trip%20can%20be%20electrified.

22. Fred Lambert, "The $4,500 Electric Car Tax Credit Bonus for Union-Made EVs is Not On The Table Anymore," Electrek, June 22, 2022, https://electrek.co/2022/06/22/electric-car-tax-credit -bonus-union-made-evs-not-on-the-table-anymore/.

23. Author interviews with people who were involved in the discussions.

24. Author interviews with people who were involved in the discussions.

25. Christine McDaniel, "The Cost Of Battery Production Tax Credits Provided In The IRA," *Forbes*, February 1, 2023, https://www.forbes.com/sites/christinemcdaniel/2023/02/01/the-cost-of -battery-production-tax-credits-provided-in-the-ira/.

26. Author interviews with people who were involved in the discussions.

27. International Energy Agency, "Inflation Reduction Act of 2022," iea.org, December 12, 2023, https://www.iea.org/policies/16156-inflation-reduction-act-of-2022.

28. Ellen Ehrenpreis, "Section 45X of the Inflation Reduction Act: New Tax Credits Available to Battery Manufacturers," Orrick law firm, November 17, 2022, https://www.orrick.com/en /Insights/2022/11/Section-45X-of-the-Inflation-Reduction-Act-New-Tax-Credits-Available-to -Battery-Manufacturers.

29. Audrey LaForest, "Why Section 45X is a 'Game Changer' for U.S. EV Battery Supply Chain," *Automotive News*, June 9, 2023, https://www.autonews.com/manufacturing/inflation -reduction-act-may-bring-billions-us-battery-makers.

30. Mike Colias, "GM Outlines Game Plan to Protect Profit in EV Transition," *Wall Street Journal*, November 17, 2022, https://www.wsj.com/articles/gm-outlines-game-plan-to-protect -profit-in-ev-transition-11668715512.

31. Institute for Energy Research, "Ford Receives a Sweet Deal From Biden to Build 3 Battery Factories," Institute for Energy Research, June 30, 2023, https://www.instituteforenergyresearch .org/regulation/ford-receives-a-sweet-deal-from-biden-to-build-3-battery-factories/#:~:text=The%20 IRA%20creates%20a%20new,totaling%20more%20than%20%247%20billion.

32. John Voelcker, "U.S.-Made EVs Could Get Massively Cheaper, Thanks to Battery Provisions in New Law," *Car and Driver*, February 3, 2023, https://www.caranddriver.com/news /a42749754/us-electric-cars-could-get-cheaper-inflation-reduction-act-section-45x/.

33. Rebecca Elliott and Mike Colias, "Tesla Shifts Battery Strategy as It Seeks U.S. Tax Credits," *Wall Street Journal*, September 14, 2022, https://www.wsj.com/articles/tesla-shifts-battery -strategy-as-it-seeks-u-s-tax-credits-11663178393.

34. Bill Schweber, "Lithium Batteries for EVs: NMC or LFP?" EETimes, January 12, 2023, https://www.eetimes.com/lithium-batteries-for-evs-go-nmc-or-lfp/.

35. Aditi Shah, "Toyota to Roll out Solid-State Battery EVs Globally in a Couple of Years," Reuters, January 11, 2024, https://www.reuters.com/business/autos-transportation/toyota-roll-out-solid-state-battery-evs-couple-years-india-executive-says-2024-01-11/.

36. Charles J. Murray, "Solid-State Batteries Could Face "Production Hell," IEEE Spectrum, January 26, 2024, https://spectrum.ieee.org/solid-state-battery-production-challenges.

37. Soyoung Kim, "A123 in Talks on Some $250 Million Loans," Reuters, October 20, 2009, https://www.reuters.com/article/retire-us-a123-idUKTRE59J4IO20091020/.

38. Karn Dhingra, "Battery Startup Our Next Energy Raises $300 Million in New Funding Found," Automotive News, February 1, 2023, https://www.autonews.com/mobility-report/our-next-energy-raises-another-300-million.

39. All Ijaz quotes and reflections in this chapter are from an author interview, August 24, 2023.

CHAPTER 14

1. Mike Colias, "Rivian Automotive Curtails Production in 2022 Due to Supply-Chain Disruptions," Wall Street Journal, March 10, 2022, https://www.wsj.com/articles/rivian-automotive-curtails-production-in-2022-due-to-supply-chain-disruptions-11646951126?mod=article_inline&mod=article_inline.

2. Jay Ramey, "Rivian Makes Progress, Still Has a Massive Backlog of Orders," AutoWeek, November 10, 2022, https://www.autoweek.com/news/green-cars/a41921710/rivian-q3-2022-financial-results/.

3. Mike Colias, "GM's Supplier-Friendly Initiatives Bear Fruit," Automotive News, August 1, 2016, https://www.autonews.com/article/20160801/OEM10/308019988/gm-s-supplier-friendly-initiatives-bear-fruit; Boeing, "Boeing Reports Fourth-Quarter Results," press release, January 31, 2024, https://s2.q4cdn.com/661678649/files/doc_financials/2023/q4/2023-12-Dec-31-8K-PR-Ex-99-1.pdf.

4. Author visit to Normal factory, November 9, 2022.

5. Author interviews with former Rivian executives and other people familiar with Scaringe's leadership style.

6. Author visit to Normal factory, November 9, 2022.

7. WSJ Video, "Inside Rivian and Ford's Plants, as They Race to Build EVs Faster," Wall Street Journal, December 22, 2022, https://www.wsj.com/video/series/in-depth-features/inside-rivian-and-fords-plants-as-they-race-to-build-evs-faster/A5F8495B-678B-4DAA-86BD-101CBA68F7FA.

8. Lucid, "Lucid Announces Third Quarter 2022 Financial Results, On Track for Annual Production Guidance of 6,000 to 7,000 Vehicles," press release, November 8, 2022,/ https://www.prnewswire.com/news-releases/lucid-announces-third-quarter-2022-financial-results-on-track-for-annual-production-guidance-of-6-000-to-7-000-vehicles-301672195.html.

9. Esha Dey, "Lucid's Market Cap Is Now Bigger Than Ford and GM," Bloomberg, November 16, 2021, https://www.bloomberg.com/news/articles/2021-11-16/lucid-overtakes-ford-gm-by-valuation-in-latest-sign-of-ev-fever?sref=PRBlrg7S.

10. Marcus Williams, "Steven David to Tackle Supply Chain Challenges at Lucid," Automotive Logistics, August. 11, 2022, https://www.automotivelogistics.media/people-and-skills/steven-david-to-tackle-supply-chain-challenges-at-lucid/43333.article#:~:text=The%20company%20is%20dealing%20with,sales%20of%20approximately%20%243.5%20billion.

11. Sean McLain, "Lucid to Raise Up To $1.5 Billion on Share Sales," Wall Street Journal, November 8, 2022, https://www.wsj.com/articles/electric-vehicle-maker-lucid-group-reports-wider-third-quarter-losses-11667944921.

12. Eliot Brown, "Electric-Vehicle Startups Promise Record-Setting Revenue Growth," *Wall Street Journal*, March 15, 2021, "https://www.wsj.com/articles/electric-vehicle-startups-promise-record-setting-revenue-growth-11615800602.

13. TRD Staff, "Fisker Founder Lists Hollywood Hills Mansion for $35M," *The Real Deal*, June 11, 2024. https://therealdeal.com/la/2024/06/11/fisker-founder-lists-hollywood-hills-mansion-for-35m/

14. Sean McLain, "A Famed Car Designer's Doomed Attempt to Challenge Tesla," *Wall Street Journal*, May 14, 2024. https://www.wsj.com/business/autos/fisker-ev-collapse-4df71216.

15. "Faraday Future Withdraws Full-Year Production Forecast As Weak EV Demand Weighs," Reuters, May 28, 2024, https://www.reuters.com/business/autos-transportation/faraday-future-withdraws-full-year-production-forecast-2024-05-28/.

16. Jamie L. LaReau, "GM Sells Its Lordstown Assembly Plant to Electric Truck Start-Up," *Detroit Free Press*, November 7, 2019, https://www.freep.com/story/money/cars/general-motors/2019/11/07/gm-lordstown-assembly-workhorse/2521887001/.

17. Peter Baker, "'People Love You': For Trump, a Welcome Escape From the Capital," *New York Times*, July 25, 2017, https://www.nytimes.com/2017/07/25/us/politics/trump-ohio-rally.html.

18. Author interviews with people familiar with GM's deliberations; Jamie L. LaReau, "How a Cleveland Car Dealer Planned to Save GM's Lordstown Plant," *Detroit Free Press*, March 19, 2019, https://www.freep.com/story/money/cars/general-motors/2019/03/19/cleveland-car-dealer-tried-save-gms-lordstown-assembly-plant/3205898002/.

19. Author interviews with people familiar with GM's deliberations.

20. Michael Wayland, "GM Sells Its Stake In Embattled EV Start-Up Lordstown Motors," CNBC, March 1, 2022, https://www.cnbc.com/2022/03/01/gm-sells-its-stake-in-embattled-ev-start-up-lordstown-motors.html.

21. Connor Smith, "Trump Says GM Will Sell Its Lordstown, Ohio Plant," Barron's, May 8, 2019.https://www.barrons.com/articles/workhorse-stock-gm-lordstown-trump-tweet-51557333676

22. "President Trump Delivers Remarks with the Lordstown Motors 2021 Endurance," YouTube, September 28, 2020, https://www.youtube.com/watch?v=WcITNzGiw2Y.

23. John Rosevear, "Lordstown Motors Says It Has Over 100,000 Reservations for Its Electric Pickup," Motley Fool, January 11, 2021, https://www.nasdaq.com/articles/lordstown-motors-says-it-has-over-100000-reservations-for-its-electric-pickup-2021-01-11.

24. Hindenburg Research, "The Lordstown Motors Mirage: Fake Orders, Undisclosed Production Hurdles, And A Prototype Inferno," March 12, 2021, https://hindenburgresearch.com/lordstown/.

25. I-Chun Chen, "Lordstown Motors Targeted by Short-Seller Hindenburg," *Cleveland Business Journal*, March 12, 2021, https://www.bizjournals.com/cleveland/news/2021/03/12/lordstown-motors-targeted-by-short-seller.html.

26. Ben Foldy and Micah Maidenberg, "Lordstown Motors Executives Resign Amid Inaccurate Preorder Disclosures," *Wall Street Journal*, June 14, 2021. https://www.wsj.com/articles/lordstown-motors-chief-executive-finance-chief-resign-11623676356.

27. Al Root, "Nikola Sells Hydrogen-Powered Trucks and the Stock Is Soaring," *Barron's*, July 13, 2023.

28. Ben Foldy, Mike Colias, and Nora Naughton, "Long Before Nikola Trucks, Trevor Milton Sold Investors on Startups That Faded," *Wall Street Journal*, October 1, 2020, https://www.wsj.com/articles/nikola-electric-hydrogen-trucks-trevor-milton-11601575695.

29. Renaissance Capital, "Electric Vehicle Startup Nikola Set to Begin Trading Under Symbol NKLA Following SPAC Merger Approval," Nasdaq.com, June 3, 2020, https://www.nasdaq.com/articles/electric-vehicle-startup-nikola-set-to-begin-trading-under-symbol-nkla-following-spac.

30. Brian Sozzi, "Nikola Founder: We Want to Crush the Ford F-150," Yahoo! Finance, June 9, 2020, https://finance.yahoo.com/news/nikola-founder-we-want-to-crush-the-ford-f-150-and-why-our-stock-is-exploding-160627980.html.

31. Author interviews with people familiar with the GM-Nikola deal.

32. "Full Interview with Nikola Chairman Trevor Milton and GM CEO Mary Barra on Partnership," cnbc.com, September 8, 2020, https://www.cnbc.com/video/2020/09/08/nikola-general -motors-partnership-full-interview-squawk-box.html.

33. Hindenburg Research, "Nikola: How to Parlay An Ocean of Lies Into a Partnership With the Largest Auto OEM in America," September 10, 2020, https://hindenburgresearch.com/nikola/.

34. Ben Foldy and Mike Colias, "Nikola's Talks With Major Energy Firms Stalled Following Short-Seller Report," *Wall Street Journal*, September 23, 2020, https://www.wsj.com/articles/nikolas -talks-with-major-energy-firms-stalled-following-short-seller-report-11600872115.

35. Hindenburg Research, Nikola, September 2020.

36. Peter Holderith, "It Would Take GMC 17 Years to Clear the Hummer EV Waitlist at Current Production Rate," *Drive*, July 1, 2022, https://www.thedrive.com/news/it-would -take-gmc-17-years-to-clear-the-hummer-ev-waitlist-at-current-production-rate.

37. Mike Colias, "GM's EV Push Stalls Amid Slow Rollouts for GMC Hummer, Cadillac Lyriq," Wall Street Journal, March 8, 2021. https://www.wsj.com/articles/gms-ev-push-stalls -amid-slow-rollouts-for-gmc-hummer-cadillac-lyriq-8c0103bd

38. Mike Colias, "Electric Vehicles Took Off. Car Makers Weren't Ready," *Wall Street Journal*, September 18, 2022, https://www.wsj.com/articles/electric-vehicles-inventory-supply-chain -batteries-11663504014.

39. Kalea Hall, "GM's EV Ramp-Up Affected By Automation Supplier Delays," Detroit News, June 25, 2023 https://www.detroitnews.com/story/business/autos/general-motors/2023/07/25 /gms-ev-ramp-up-affected-by-automation-supplier-delays/70461987007/.

40. "'Production And Logistics Hell'—Elon Musk Said Tesla Was Close To Bankruptcy," CBS News, November 4, 2020, https://www.cbsnews.com/sanfrancisco/news/elon-musk-tesla-close -to-bankruptcy/.

41. Matt Hardigree, "The Tesla Cybertruck Is Going To Be So Hard To Build," *Autopian*, November 28, 2023, https://www.theautopian.com/the-tesla-cybertruck-is-going-to-be-so -hard-to-build/.

42. Author interview, August 29, 2023.

43. Author interview, August 29, 2023.

44. Author interview with a former Rivian executive who worked closely with Scaringe, February 3, 2023.

45. Author interview with a former Rivian executive who worked closely with Scaringe, February 3, 2023.

46. Author interview, August 29, 2023.

47. Author interview, August 29, 2023.

48. Author interview, August 29, 2023.

49. Jay Ramey, "Update: Rivian Reverses Sudden Price Hikes," AutoWeek, March 2, 2022 https://www.autoweek.com/news/green-cars/a39295674/rivian-quad-motor-price-hikes/.

50. Author interview, July 2023.

CHAPTER 15

1. Author interview, November 28, 2023.

2. Cox Automotive, "New-Vehicle Inventory, Prices Stabilize; EV Supply Grows," press release, July 13, 2023, https://www.coxautoinc.com/market-insights/new-vehicle-inventory-june-2023/.

3. Chris Lemley, LinkedIn, https://www.linkedin.com/in/lemley/.

4. Author interview, November 28, 2023.

5. Author interview, August 8, 2023.

6. Michelle Lewis, "Here's What's Projected for Electric Vehicle Manufacturing Through 2030," Electrek, April 12, 2022, https://electrek.co/2022/04/12/heres-whats-projected-for-electric -vehicle-manufacturing-through-2030/#:~:text=In%20the%20US%20alone%2C%20 13,cars%2C%20trucks%2C%20and%20SUVs.

7. Christopher Ludwig, "The Future of Ford in a Flash of Lightning," Automotive Manufacturing Solutions, October 19, 2023, https://www.automotivemanufacturingsolutions.com /factory-expansion/the-future-of-ford-in-a-flash-of-lightning/44498.article.

8. Kai Ryssdal and Sean McHenry, "Scaling EVs Means Skipping 'Half steps' with Hybrids, GM's Mary Barra Says," Marketplace, June 8, 2023, https://www.marketplace.org/2023/06/08 /gm-ceo-mary-barra-electric-vehicles-charging-stations/.

9. Carrie Hampel, "Volkswagen Electric Car Deliveries Are Up in 2022," electdrive, October 15, 2022, https://www.electrive.com/2022/10/15/volkswagen-electric-car-deliveries-are-up-in -2022/.

10. Mike Colias, "Electric Vehicles Took Off. Car Makers Weren't Ready," *Wall Street Journal*, September 18, 2022, https://www.wsj.com/articles/electric-vehicles-inventory-supply-chain -batteries-11663504014.

11. Author interview, January 5, 2024.

12. Marianne Simmons, Twitter post, January 12, 2023, https://x.com/MarianneSimmons /status/1613754758598205442?lang=en.

13. Motley Fool Transcribing, "Tesla (TSLA) Q4 2022 Earnings Call Transcript," January 26, 2023, https://www.fool.com/earnings/call-transcripts/2023/01/26/tesla-tsla-q4-2022-earnings -call-transcript/.

14. Author interview with GM executive on condition of anonymity, April 2023.

15. Author interview, January 4, 2024.

16. "Path to EV Adoption: Consumer and Dealer Perspectives," Cox Automotive, June 2023, https://www.coxautoinc.com/wp-content/uploads/2023/06/Path-to-EV-Adoption-Study -Summary-June-2023.pdf.

17. "Inflation Reduction Act of 2022," Ballotpedia, https://ballotpedia.org/Inflation _Reduction_Act_of_2022.

18. David Shepardson, "Republican Takeover of U.S. House Could Pump the Brakes on Biden's EV Agenda," Reuters, November 18, 2022, https://ca.finance.yahoo.com/news/republican -takeover-u-house-could-160157342.html.

19. Mike Colias, "Another Roadblock to the EV Transition: Personal Politics," *Wall Street Journal*, May 27, 2024, https://www.wsj.com/business/autos/another-roadblock-to-the-ev -transition-personal-politics-ab4e311b.

20. Neal E. Boudette, "For Bill Ford, 'Every Negotiation Is a Roller Coaster,'" *New York Times*, October 18, 2023, https://www.nytimes.com/2023/10/18/business/bill-ford-corner -office.html.

21. Copy of letter to President Biden from more than 4,000 car dealers, November 2023, https://44308654.fs1.hubspotusercontent-na1.net/hubfs/44308654/EV%20Letter%201.pdf.

22. Mike Colias, "Mary Barra Spent a Decade Transforming GM. It Hasn't Been Enough," *Wall Street Journal*, December 16, 2023, https://www.wsj.com/business/autos/mary-barra -spent-a-decade-transforming-gm-it-hasnt-been-enough-d82f4c5a.

23. Alexa St. John and Nora Naughton, "Auto Execs Are Coming Clean: EVs Aren't Working," *Business Insider*, October 26, 2023, https://www.businessinsider.com/auto-executives-coming -clean-evs-arent-working-2023-10.

24. Ford Motor Company, Third Quarter 2023 Earnings Conference Call, October 26, 2023. https://s201.q4cdn.com/693218008/files/doc_financials/2023/q3/Ford-Q3-2023-Earnings-Call -Transcript.pdf.

25. Gabrielle Coppola, "Battery Startup ONE Demotes Founder and CEO During Cash Crunch," Bloomberg, December 10, 2023, https://www.bloomberg.com/news/articles/2023-12-10 /battery-startup-one-demotes-founder-and-ceo-during-cash-crunch?sref=PRBlrg7S.

26. Tesla (TSLA), Q3 2023 Earnings Call Transcript, October 18, 2023, https://www.fool.com /earnings/call-transcripts/2023/10/18/tesla-tsla-q3-2023-earnings-call-transcript/.

27. "S&P Global Mobility Survey Finds EV Affordability tops Charging and Range Concerns in Slowing EV Demand," S&P Global Mobility, November 8, 2023, https://press.spglobal.com /2023-11-08-S-P-Global-Mobility-Survey-Finds-EV-Affordability-tops-Charging-and-Range -Concerns-in-Slowing-EV-Demand#:~:text=Almost%20half%20(48%25)%20of,analyst%20at%20 S%26P%20Global%20Mobility.

28. Ford Motor Company, 2023 Fourth Quarter Financial Results, February 6, 2024, https://s201 .q4cdn.com/693218008/files/doc_financials/2023/q4/Ford-Q4-2023-Earnings-Call-Transcript.pdf.

29. Tesla (TSLA), Q3 2023 Earnings Call Transcript, October 18, 2023.

30. "R2, R3, R3X Revealed–Rivian," YouTube, https://www.youtube.com/watch?v=82IiXIKkRIA.

31. Peter Valdes-Depena, "A Chinese EV Maker Just Revealed a 1,300 Horsepower Supercar," CNN, February 26, 2024, https://www.cnn.com/2024/02/26/cars/chinese-ev-maker-byd-supercar /index.html.

32. Burkhard Riering, "How BYD's Giant China Operation Drives the Automaker's Global Expansion," Automobilwoche, April 27, 2024, https://europe.autonews.com/automakers/byds -global-ev-push-driven-giant-china-operation.

33. Akiko Fujita, "BYD Is Not Interested in United States' EV Market: Exec. VP," Yahoo! Finance, February 26, 2024, https://finance.yahoo.com/video/byd-not-interested-united-states -182445977.html.

CONCLUSION

1. International Energy Agency, "Global EV Outlook 2024: Trends in Electric Cars," IEA website. https://www.iea.org/reports/global-ev-outlook-2024/trends-in-electric-cars

2. "Norwegian EV Market Surges to 91.5% Market Share, Setting a Sustainable Example," European Commission's European Alternative Fuels Observatory, April 7, 2024, https:// alternative-fuels-observatory.ec.europa.eu/general-information/news/norwegian-ev-market -surges-915-market-share-setting-sustainable-example.

3. Jamie L. LaReau, "GM's Cadillac Backtracks, Says Gas Vehicles Likely to Stay in Lineup Beyond 2030," Detroit Free Press, May 1, 2024, https://www.freep.com/story/money/cars/general -motors/2024/05/01/gm-cadillac-brand-internal-combustion-engine-gas-vehicles-electric-2030 /73527130007/.

4. Peter Johnson, "Mercedes-Benz Drastically Backtracks EV Plans, Will Build Gas Cars Well Into 2030s," Electrek, February 22, 2024, https://electrek.co/2024/02/22/mercedes-backtracks -ev-plans-gas-cars-2030s/#:~:text=Mercedes%20backtracks%20on%20EV%2Donly%20 commitment%20by%202030&text=The%20luxury%20automaker%20said%20 all,and%20the%20EQB%20electric%20SUV.

5. "Survey Highlights Americans' Continued Interest in EVs," EV Report, June 6, 2024, https://theevreport.com/survey-highlights-americans-continued-interest-in-evs.

6. Umar Shakir, "Ram's New 1500 EV Truck has 'Unlimited' Range, Thanks to Built-In Gas Generator," The Verge, November 7, 2023.

7. Danny Lee, "New BYD Hybrid Can Drive Non-Stop for More Than 2,000 Kilometers," Bloomberg via Yahoo!, May 28, 2024, https://finance.yahoo.com/news/byd-shows-off-hybrid -powertrain-125608793.html.

8. Author interview, August 8, 2023.

9. Abhirup Roy, "Tesla CEO Musk: Chinese EV Firms Will 'Demolish' Rivals Without Trade Barriers," Reuters, January 25, 2024, https://www.reuters.com/business/autos-transportation/tesla-ceo-musk-chinese-ev-firms-will-demolish-rivals-without-trade-barriers-2024-01-25/.

10. Gregor Sebastian, Noah Barkin and Agatha Kratz, "Ain't No Duty High Enough," Rhodium Group, April 29, 2024, https://rhg.com/research/aint-no-duty-high-enough/.

11. Laure He, "Volkswagen Invests $700 Million in Chinese EV Maker Xpeng to Boost Sluggish Sales," CNN, July 27, 2023, https://www.cnn.com/2023/07/27/cars/china-volkswagen-xpeng-investment-intl-hnk/index.html.

12. Arjun Kharpal, "Jeep, Dodge Maker Stellantis to Invest $1.6 Billion in Chinese EV Startup Leapmotor," CNBC, October 26, 2023, https://www.cnbc.com/2023/10/26/stellantis-to-invest-1point6-billion-in-chinese-ev-start-up-leapmotor.html.

13. Author interview, June 14, 2024.

INDEX

TK

ACKNOWLEDGMENTS

This book exists only because more than a hundred people trusted me to tell their stories. Thanks to every interviewee who committed their time. I owe a special debt of gratitude to my *Wall Street Journal* colleague and former Detroit newsroom neighbor Ben Foldy, who was integral in conceiving this project, before he motored on from the autos beat to cover financial crime in New York.

My editor for this book, *Harvard Business Review*'s Scott Berinato, showed much thoughtfulness and patience with me, a first-time author. His deft touch and vision helped string together my collection of disparate anecdotes and scenes into something of a coherent narrative. The diligence of fact-checker Roni Greenwood from Girl Friday Productions was much appreciated.

The foundation for this book was rooted in years of reporting on the EV transition for the *Journal*, where I've had the honor to work for eight years. Thanks to John Stoll for bringing me into the Journal fold. As Detroit bureau chief, my longtime colleague and friend Christina Rogers was instrumental in overseeing hundreds of individual stories over the years—written by me and many talented colleagues—which helped shape this broader narrative. A big thank-you also to the *Journal*'s Editor in Chief Emma Tucker; business editor, Jamie Heller, and deputy corporate coverage chief George Stahl, whose guidance and encouragement allowed us to pursue this important story from many angles.

My introduction to the auto industry came during an invaluable six years as a reporter at *Automotive News*, where I learned so much from so many, including Jason Stein—who took a chance on a car-biz newbie—Dave Versical, Krishnan Anantharaman, Phil Nussel, Jim Treece, Rick Johnson, Brad Wernle, and Larry Vellequette.

As a rookie author, I relied on many people for advice throughout this process. My agent, Ryan Harbage, provided spot-on direction during the conception and proposal writing. Several *Journal* colleagues shared their experiences and tips from their own first books, including Tim Higgins, Sean McLain, and Peter Loftus. And thanks to my whip-smart niece Reagan Snyder for her research help.

Finally, a project like this is only possible with the loving support of family and friends. A huge thank-you to my wife, Stacy, and three boys, Nico, Toby, and Teddy, for putting up with many nights, weekends, and vacations with me tucked into my laptop.